AIAA
G-045-2003

Guide

Assessing Experimental Uncertainty — Supplement to AIAA S-071A-1999

Sponsored by

American Institute of Aeronautics and Astronautics

Abstract

This AIAA Guide supplements the methodology for assessment of experimental uncertainties and techniques for evaluating experimental error sources provided in AIAA Standard S-071A-1999, "Assessment of Experimental Uncertainty with Application to Wind Tunnel Testing." This document provides additional information and examples to assist the experimentalist in performing an uncertainty analysis. Its focus is on helping the experimenter begin to apply uncertainty analysis techniques. The information contained in the standard and this guide is not limited to wind tunnel testing—it can be applied to a wide range of experiments.

Library of Congress Cataloging-in-Publication Data

Guide : assessing experimental uncertainty : supplement to AIAA S-071A-1999 / sponsored by American Institute of Aeronautics and Astronautics.
   p. cm.
Includes bibliographical references.
  ISBN 1-56347-663-0 (hardcopy) -- ISBN 1-56347-664-9 (electronic)
  1. Wind tunnels. 2. Airplanes--Models--Testing--Standards. 3. Uncertainty (Information theory) I. Title: Assessing experimental uncertainty. II. American Institute of Aeronautics and Astronautics. III. Assessment of experimental uncertainty with application to wind tunnel testing.
  TL567.W5A87 1999 Suppl.
  629.134'52--dc22

2003024006

Published by
American Institute of Aeronautics and Astronautics
1801 Alexander Bell Drive, Suite 500, Reston, VA 20191-4344

Copyright © 2003 American Institute of Aeronautics and Astronautics
All rights reserved.

No part of this publication may be reproduced in any form, in an electronic retrieval system or otherwise, without prior written permission of the publisher.

Printed in the United State of America

CONTENTS

Foreword v
Dedication vi
Part I   Basic Topics 1
1   Introduction 1
2   Uncertainty Methodology and Application 2
    2.1   Methodology Primer 2
    2.1.1   Single Test 4
    2.1.2   Multiple Tests 4
    2.1.3   Relative Sensitivity Factors 5
    2.2   Application Primer 5
    2.2.1   Experiment Definition 5
    2.2.2   Uncertainties of the Experimental Results 6
    2.2.2.1   Multiple Tests 6
    2.2.2.2   Single Test 9
3   General Uncertainty Analysis Example—Evaluating Measurement Methods 11
    3.1   Introduction 11
    3.2   Mach Number Equations 11
    3.3   Uncertainty Analysis 12
    3.3.1   Partial Derivative Terms 13
    3.3.2   Transducer Uncertainty 14
    3.3.3   Mach Number Uncertainty 14
    3.3.4   Relative Sensitivity Factors 16
4   General Uncertainty Analysis Example—Evaluating Data Reduction Equations 18
    4.1   Turbine Efficiency Equations 19
    4.2   Procedure 19
    4.3   Results 20
    4.3.1   Overall Uncertainty 20
    4.3.2   UMF 20
    4.3.3   UPC 21
    4.3.4   Summary 22
5   Systematic Uncertainties and Correlation 30
    5.1   Systematic Uncertainties 30
    5.2   Correlated Systematic Uncertainties 32
    5.2.1   Bath Uniformity 33
    5.2.2   Standard Uncertainty 34

5.3	Correlated Systematic Uncertainties Example	34
Part II	Advanced Topics	38
6	Random Uncertainties and Correlation	38
6.1	Random Uncertainties	38
6.2	Correlated Random Uncertainty Example	38
6.2.1	Background	38
6.2.2	Calibration Approach and Results	39
6.2.3	Uncertainty Analysis	40
6.3	Discharge Coefficient and Mass Flow Rate Equations	44
7	Regression Uncertainty	47
7.1	Categories of Regression Uncertainty	47
7.1.1	Uncertainty in Coefficients	48
7.1.2	Uncertainty in Y from Regression Model	48
7.1.3	(X_i, Y_i) Variables are Functions	49
7.2	Linear Regression Uncertainty	49
7.2.1	General Approach	49
7.2.2	Reporting Regression Uncertainties	51
7.2.3	Differential Pressure Transducer Calibration Example	52
7.2.4	X and Y as Functional Relations	54
7.2.5	1st Order Regressions	56
7.2.5.1	Uncertainty in Coefficients	56
7.2.5.2	Classical Regression Random Uncertainty	58
7.2.5.3	Lift Slope and Lift Coefficient Example	59
8	Automated Uncertainty Analysis for Production Experiments	64
8.1	Introduction	64
8.2	Error Propagation	64
8.3	Estimation of Input Uncertainties	65
8.3.1	Systematic Uncertainty Estimation	66
8.3.2	Random Uncertainty Estimation	67
8.3.3	Specific Uncertainty Estimates	67
8.4	Results	68
8.5	Summary	68
9	Useful References	79

Foreword

Experimental uncertainty is a complex subject involving both statistical techniques and engineering judgment. In 1995 the AIAA Standards Technical Council approved an AIAA Standard on experimental uncertainty that was revised in 1999 (AIAA S-071A-1999, "Assessment of Experimental Uncertainty with Application to Wind Tunnel Testing"). The AIAA adopted the contents of an Advisory Report (AR-304) set forth by the Advisory Group for Aerospace Research and Development (AGARD). The AIAA Standard defines a rational and practical framework for quantifying and reporting experimental uncertainty and presents the application of the methodology to wind tunnel testing. This AIAA Guide supplements the standard. It provides additional information and examples to assist the experimentalist in performing an uncertainty analysis. Its focus is on helping the experimenter begin to apply uncertainty analysis techniques.

The AIAA Standards Procedures provide that all approved Standards, Recommended Practices, and Guides are advisory only. Their use by anyone engaged in industry or trade is entirely voluntary. There is no agreement to adhere to any AIAA standards publication and no commitment to conform to or be guided by a standards report. In formulating, revising, and approving standards publications, the Committees on Standards will not consider patents that may apply to the subject matter. Prospective users of the publications are responsible for protecting themselves against liability for infringement of patents or copyrights, or both.

The Standards Subcommittee of the AIAA Ground Test Technical Committee (Mr. E.A. Arrington, Chairperson) had many members who contributed to this document over the course of its preparation. The following individuals should be recognized for their contributions:

Gregory Addington, Chairperson	Air Force Research Lab
David M. Cahill	Sverdrup Tech., AEDC
Julie Carlile	Air Force Research Lab
Daniel Cresci	GASL
Susan T. Hudson, Technical Editor-in-Chief	Mississippi State University
Wayne Kalliomaa	Air Force Research Lab
Jerry Kegelman	NASA Langley Research Center
Daniel E. Marren	AEDC/White Oak
Laura J. McGill	Raytheon
Thomas McLaughlin	U.S. Air Force Academy
Julie Morrow	U.S. Air Force Academy
Mathew L. Rueger	Boeing
William A. Straka	Applied Research Lab., Penn State University
James C. Yu	NASA Langley Research Center

The committee members would like to acknowledge the following individuals for their valuable contributions and reviews of this document:

Kendall Brown	NASA Marshall Spaceflight Center
Mark Kammeyer	Boeing
Hugh Coleman	Univ. of Alabama in Huntsville
Glenn Steele	Mississippi State University
Stephen McClain	University of Alabama in Birmingham

The principal authors and reviewers of this document were: Susan Hudson, Kendall Brown, Mark Kammeyer, David Cahill, Hugh Coleman, Glenn Steele, and Stephen McClain.

The AIAA Ground Test Technical Committee (Mr. Allen Arrington, Chairman) approved the document for publication in January 2003

The AIAA Standards Executive Council (Mr. Phil Cheney, Chairman) accepted the document for publication in September 2003.

The AIAA Standards Procedures provide that all approved Standards, Recommended Practices, and Guides are advisory only. Their use by anyone engaged in industry or trade is entirely voluntary. There is no agreement to adhere to any AIAA standards publication and no commitment to conform to or be guided by a standards report. In formulating, revising, and approving standards publications, the Committees on Standards will not consider patents which may apply to the subject matter. Prospective users of the publications are responsible for protecting themselves against liability for infringement of patents or copyrights, or both.

Dedication

This document is dedicated to the late Frank Wright of Boeing Commercial Airplane Company. Frank served as an outstanding role model for those practicing experimental uncertainty analysis and its applications to data quality assurance in ground testing. He was a world-wide crusader for experimental uncertainty analysis. Most of us on the Standards Subcommittee had the privilege of working with Frank. It was his inspiration and dedication to excellence in wind tunnel testing in general and data quality in particular that started this project. Frank developed the first draft of the guide in 1996 when he was the Chair of the Standards Subcommittee. Much of Frank's vision is incorporated in this Guide. Our hope is that, through this Guide, more engineers will be inspired as we were inspired by Frank.

Part I Basic Topics

1   Introduction

Experimental uncertainty has long been a topic of discussion and controversy in the aerospace community. The problem is not the lack of good methodology in this area; references on the subject exist and are readily available. The difficulty has been in the application of the methodology with consistency and regularity. For the novice, or even the experienced researcher, the mass of equations presented in the references can be intimidating. Further, although some examples illustrating the application of the uncertainty analysis methodology are presented in the references, it is sometimes difficult for the reader to grasp how the examples apply to their particular testing requirements. This document provides supplemental information and examples to assist the experimentalist in performing an uncertainty analysis. Its focus is on helping one get started. This document assumes that the reader has reviewed the Standard to become familiar with the principles and terminology. The examples given are simply meant to guide one through the uncertainty analysis process for particular cases so that the process can be more easily understood. They are not meant to define techniques for conducting similar experiments, though sound engineering principles are applied throughout the document.

Uncertainty analysis is a useful and essential part of an experimental program. Data quality assessment is a key part of the entire testing process, and should, therefore, be applied in all phases of an experiment. A general analysis performed in the planning phase allows a comprehensive examination of the experimental process. This type of analysis can inform one beforehand, during the design of the experiment, when the experiment cannot meet the desired uncertainty limits. It can also show when improved instruments and/or improved processes must be found to obtain a given output uncertainty. Similarly, it can identify instrumentation and/or processes that control uncertainty and allow changes to be made to meet test goals. Additionally, a general uncertainty analysis can provide a check against unknowingly taking data under test conditions where uncertainties become intolerably large.

Once the test data is collected, a detailed uncertainty analysis should be performed. The detailed analysis uses the actual experimental data and the uncertainties associated with this data. This analysis provides an assessment of the quality of the experimental data and provides a basis for comparing the test results with analyses or results from other test facilities.

Any uncertainty analysis requires "engineering judgment." A lack of explicit information for uncertainty estimates does not prohibit performing an uncertainty analysis. The use of some appropriate uncertainty analysis is indispensable in experimentation and any appropriate or approximate analysis is far better than no analysis, as long as the process is explained for the user.

This document will provide examples to help with the application of uncertainty analyses. It will begin with a brief overview of the methodology. (Details of the methodology are contained in the Standard.) This will be followed by a detailed example of applying uncertainty analysis to a simple situation to illustrate the basics of the methodology. Increasingly complex examples will then address various uncertainty analysis aspects. Details of the calculations will be shown in the early chapters to aid in understanding. As the examples become more complex and the reader becomes more familiar with the methodology, fewer calculations will be shown. A comprehensive example for an experimental system is given as the last example to further demonstrate the potential of uncertainty analysis. The document will end with a modern list of useful references. Since the field of uncertainty analysis is constantly evolving, one must be particularly careful to use up-to-date references.

2 Uncertainty Methodology and Application

2.1 Methodology Primer

This section will give a brief overview of uncertainty analysis methods and how errors propagate through a given data reduction equation. A comprehensive discussion of uncertainty analysis techniques is contained in the AIAA Uncertainty Standard (Ref. 2.1). The reader is directed to the Standard and reference 2.2 to get in-depth information on experimental uncertainty and its application to an experimental process.

The word accuracy is generally used to indicate the relative closeness of agreement between an experimentally-determined value of a quantity and its true value. Error is the difference between the experimentally-determined value and its true value; therefore, as error decreases, accuracy is said to increase. Since the true value is not known, it is necessary to estimate error, and that estimate is called an uncertainty, U. Uncertainty estimates are made at some confidence level—a 95% confidence estimate, for example, means that the true value of the quantity is expected to be within the ±U interval about the experimentally-determined value 95 times out of 100.

Total error can be considered to be composed of two components: a random (precision) component, ε, and a systematic (bias) component, β. The terms precision and random as well as bias and systematic will be used interchangeably in this document. The classification of errors as random or systematic is defined in the Standard and maintained in this Guide. An error is classified as random if it contributes to the scatter of the data; otherwise, it is a systematic error. As an estimator of β, a systematic uncertainty or bias limit, B, is defined. A 95% confidence estimate is interpreted as the experimenter being 95% confident that the true value of the systematic error, if known, would fall within ±B. A useful approach to estimating the magnitude of a systematic error is to assume that the systematic error for a given case is a single realization drawn from some statistical parent distribution of possible systematic errors. Or, in other words, the systematic error could be treated as a random variable, but with only a single realization, its variance cannot be measured and must be estimated. As an estimator of the magnitude of the random errors, a random uncertainty or precision limit, P, for a single reading is defined. A 95% confidence estimate of P is interpreted to mean that the ±P interval about the single reading of X_i should cover the (biased) parent population mean, μ, 95 times out of 100.

In nearly all experiments, the measured values of different independent variables are combined using a data reduction equation (DRE) to form some desired result. A general representation of a DRE is

$$r = r(X_1, X_2, ..., X_J) \tag{2.1}$$

where r is the experimental result determined from J independent variables X_i. Each of the measured variables contains systematic errors and random errors. These errors then propagate through the DRE, thereby generating the systematic and random errors in the experimental result, r.

The 95% confidence expression for U_r is

$$U_r^2 = B_r^2 + P_r^2 \tag{2.2}$$

If it is assumed that the degrees of freedom for the result is large (>10), which is very appropriate for most engineering applications, then the "large sample assumption" applies [Ref. 2.2] and the 95% confidence expression for U_r is

$$B_r^2 = \sum_{i=1}^{J} \theta_i^2 B_i^2 + 2 \sum_{i=1}^{J-1} \sum_{k=i+1}^{J} \theta_i \theta_k B_{ik} \tag{2.3}$$

where

$$\theta_i = \frac{\partial r}{\partial X_i} \qquad (2.4)$$

The bias limit estimate for each X_i variable is the root sum square combination of its elemental systematic uncertainties

$$B_i = \left[\sum_{j=1}^{M}(B_i)_j^2\right]^{1/2} \qquad (2.5)$$

where M is the number of elemental systematic errors.

B_{ik}, the 95% confidence estimate of the covariance appropriate for the systematic errors in X_i and X_k, is determined from

$$B_{ik} = \sum_{\alpha=1}^{L}(B_i)_\alpha (B_k)_\alpha \qquad (2.6)$$

where variables X_i and X_k share L identical error sources. These terms account for correlation between <u>systematic errors</u> in different measurements. More detailed discussions of the covariance approximation are given in Ref. 2.2 and 2.3.

The random uncertainty (precision limit) of the result is

$$P_r^2 = \sum_{i=1}^{J}\theta_i^2 P_i^2 + 2\sum_{i=1}^{J-1}\sum_{k=i+1}^{J}\theta_i\theta_k P_{ik} \qquad (2.7)$$

where P_{ik} is the 95% confidence estimate of the covariance appropriate for the random errors in X_i and X_k. The 95% confidence large sample (number of readings, N≥10) precision limit for a variable is estimated as

$$P_i = 2S_i \qquad (2.8)$$

The sample standard deviation for X_i is

$$S_i = \left[\frac{1}{N-1}\sum_{k=1}^{N}[(X_i)_k - \overline{X}_i]^2\right]^{1/2} \qquad (2.9)$$

where N is the number of measurements and the mean value for X_i is defined as

$$\overline{X}_i = \frac{1}{N}\left[\sum_{k=1}^{N}(X_i)_k\right] \qquad (2.10)$$

The 95% confidence large sample precision limit for a mean value of X_i is estimated as

$$P_{\overline{X}_i} = \frac{2S_{\overline{X}_i}}{\sqrt{N}} \qquad (2.11)$$

Equation 2.11 assumes a sufficient time interval for the N measurements such that each one can be considered independent. Typically, correlated random uncertainties have been neglected so that the P_{ik}'s in Eq. 2.7 are taken as zero. These covariance terms account for correlation between <u>errors</u> in different measurements. The random errors have been considered to be random; therefore, the correlation between them has been assumed to be zero. That assumption is often true; however, a case is presented in Chapter 6 where the random errors are correlated and the covariance terms are important. These terms will be discussed further in Chapter 6.

It is important that the uncertainty analysis be performed using a DRE that is in a form that contains only independent parameters. The partial derivatives must be taken with respect to the independent parameters. A good example of the consequences that can result from using partial derivatives taken with respect to a derived quantity is illustrated in Annex 4C of the Standard.

Occasionally, a DRE is simple enough that the partial derivatives can be taken by hand (see the examples following in this chapter). However, the DRE's are often complex and determining the partial derivatives by hand is difficult. Fortunately, there are several methods that can be used to determine the partial derivatives. A common method of determining the partial derivatives is through the use of a symbolic mathematical program such as Mathematica®. Alternately, the partial derivatives can be determined numerically using "jitter" programs (Ref. 2.2). The jitter program can be written in various programming languages for individual applications or software packages such as MathCad® can be used. The most recent development for obtaining the partial derivatives are the automatic differential programs such as ADIFOR® (Ref. 2.4). These programs can be used with the data reduction program to automatically determine the partial derivatives. It may also be possible to combine the automatic differential programs with other methods (i.e., parameter overloading) to automatically produce the desired uncertainties based on the inputs provided to the program (Ref. 2.5).

2.1.1 Single Test

The precision limits defined in Eqs. 2.8 and 2.11 and used in Eq. 2.7 are applicable to a single test with single readings or average values from multiple readings of the same variable. For a single test, the result is determined once at a given test condition using Eq. 2.1. The situation in which Eq. 2.8 applies is often encountered in engineering tests and occurs when measurements of the variables are averaged over a period of time that is small compared to the time periods of the factors causing variability in the experiment. A proper precision limit cannot be calculated from readings taken over such a small time interval (Ref. 2.2). For such data, the measurement(s) of a variable should be considered a single reading, and the precision limit must be estimated based on previous information about that measurement obtained over the appropriate time interval. If previous readings of a variable over an appropriate interval are not available, then the experimenter must estimate a value for P_i using the best information available at the time. (Calibration data, previous testing in the same facility, previous testing using similar equipment, etc., can be used to estimate P_i.) If multiple readings of a single variable can be made over an appropriate time interval, then Eq. 2.11 is applicable.

2.1.2 Multiple Tests

If a test is repeated a number of times so that multiple results at the same test condition are available, then the best estimate of the result r would be $\bar{r}$.

$$\overline{X}_i = \frac{1}{m}\left[\sum_{k=1}^{m}(X_i)_k\right] \qquad (2.12)$$

m is the number of separate test results. The precision limit for this result would be $P_{\bar{r}}$ calculated as

$$P_{\bar{r}} = \frac{K S_r}{\sqrt{m}} \qquad (2.13)$$

K is the coverage factor and is taken as 2 for large sample sizes (Ref 2.1 and 2.2). As before, S_r is the standard deviation of the sample of m results and is defined as

$$S_r = \left[\frac{1}{m-1}\sum_{k=1}^{m}[r_k - \bar{r}]^2\right]^{1/2} \qquad (2.14)$$

Obviously, this cannot be computed until multiple results are obtained. Also note that the precision limit computed is only applicable for those random error sources that were "active" during the repeat measurements. For example, if the test conditions were not changed and then reestablished between the multiple results, the variability due to resetting to a given test condition would not be accounted for in the precision estimate. Note that the approach described in this section implicitly includes the correlated precision effect (see Chapter 6).

2.1.3 Relative Sensitivity Factors

It is very useful to investigate the sensitivity of the test result uncertainty to the uncertainty of the various measured quantities. When used for a general uncertainty analysis, this allows one to concentrate efforts on critical measurements and make changes to meet the test goals. An example of using sensitivity factors in a general analysis will be given in Chapter 4. When used in the detailed analysis, this allows one to see how the measurements actually influenced the uncertainty of the final result. Sensitivity factors are presented in two ways: the uncertainty magnification factor (UMF) and the uncertainty percentage contribution (UPC). (Ref. 2.2, 2.6, 2.7)

The UMF is defined as

$$UMF_i = \frac{X_i}{r}\theta_i = \frac{X_i}{r}\frac{\partial r}{\partial X_i} \tag{2.15}$$

The UMF illustrates the influence of the uncertainty of one variable as it propagates through the DRE into the result if the uncertainties of all other variables are zero. If the UMF is less than one, the variable uncertainty diminishes as it propagates through the DRE or DRE; if the UMF is greater than 1, the variable uncertainty increases as it propagates through the DRE. The UMF simply evaluates the DRE—it is independent of the facility or uncertainty estimates. This analysis can be done very early in the test program before details of the testing are known.

The UPC is defined as

$$UPC_i = \frac{(\theta_i U_i)^2}{U_r^2}*100 = \frac{\left(\frac{X_i}{r}\frac{\partial r}{\partial X_i}\right)^2 \left(\frac{U_{X_i}}{X_i}\right)^2}{\left(\frac{U_r}{r}\right)^2}*100 \tag{2.16}$$

$$UPC_{ik} = \frac{(2\theta_i \theta_k U_{ik})}{U_r^2}*100 \tag{2.17}$$

where U can be systematic, random, or overall uncertainty, depending on the type of analysis being performed. The significance of the UPC terms can be seen by referring to Eqs. 2.3 and 2.7. The UPC illustrates the influence of each variable and its uncertainty as a percent of the result uncertainty squared for each squared term. The sum of the contributions of all of the UPC terms is 100%. This approach shows the sensitivity of the squared uncertainty of the result to the squared uncertainty effect of each of the variables for a particular situation where values for the variables are known and the uncertainties for each variable have been estimated. The UPC analysis considers the uncertainty of each variable and, therefore, extends the UMF analysis to a particular facility or test. Since this type of analysis is an extension of the UMF analysis that incorporates the uncertainty estimates associated with a particular test situation, it is useful from the later planning phase throughout an experiment. The uncertainty estimates should be realistic estimates based on the best information available at the time of the analysis. (Note that this was called the "uncertainty sensitivity percent" or "USP" in Ref. 2.6 and modified to "UPC" in Ref. 2.7 and 2.2)

2.2 Application Primer

2.2.1 Experiment Definition

Now that an overview of the uncertainty methodology has been presented, it will be applied to two simple examples. First, consider a case where the pressure difference between two points is needed (Experiment 1). Two pressure measurements are made: one pressure is measured at plane 1 and one pressure is measured at plane 2 (Fig. 2.1a). The required result, delta pressure, is then calculated using the following DRE:

$$\Delta P = P_1 - P_2 \qquad (2.18)$$

Second, consider that an pressure at a cross-section is needed (Experiment 2). Three pressure measurements are made at this cross-section (Figure 2.1b). The three pressures are then averaged to calculate the pressure at the cross-section, P_{CS}, using the following DRE:

$$P_{CS} = (P_3 + P_4 + P_5)/3 \qquad (2.19)$$

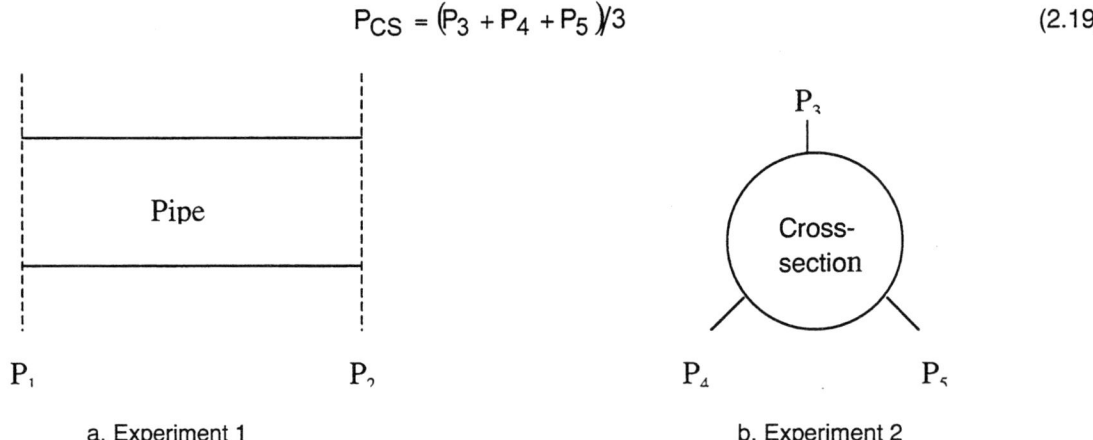

a. Experiment 1   b. Experiment 2

Figure 2.1   Experiment Representations

2.2.2 Uncertainties of the Experimental Results

The uncertainty methodology presented in section 2.1 will now be applied to these two examples for both multiple test and single test methodologies. The analysis will be a detailed uncertainty analysis where systematic and random terms are considered separately.

2.2.2.1 Multiple Tests

In a multiple test experiment, multiple data points are taken so that multiple results can be calculated. The data points have been taken over an appropriate time interval so that the multiple results calculated from these data points can be considered independent. The pressure measurements and results for the two experiments here are shown in Table 2.1.

Table 2.1 — Results from Experiments 1 and 2

Data Point	P_1 (psia)	P_2 (psia)	ΔP	P_3 (psia)	P_4 (psia)	P_5 (psia)	P_{CS}
1	12.003	7.015	4.988	9.535	9.600	9.462	9.532
2	11.880	6.890	4.990	9.583	9.556	9.593	9.577
3	11.999	7.035	4.964	9.479	9.515	9.577	9.524
4	12.037	7.013	5.024	9.515	9.479	9.402	9.465
5	12.009	6.903	5.106	9.409	9.462	9.469	9.447
6	11.794	6.999	4.795	9.495	9.480	9.483	9.486
7	12.020	7.029	4.991	9.590	9.572	9.583	9.582
8	11.997	6.803	5.194	9.515	9.591	9.579	9.562
9	12.020	7.030	4.990	9.421	9.437	9.401	9.420
10	12.003	6.799	5.204	9.399	9.572	9.599	9.523
Mean	11.976	6.952	5.025	9.494	9.526	9.515	9.512
S	0.077	0.095	0.120	0.068	0.059	0.080	0.056
2S	0.154	0.189	0.239	0.136	0.118	0.160	0.111

From Table 2.1, the average experimental results are

$$\overline{\Delta P} = 5.025 \text{ psi} \tag{2.20}$$

$$\overline{P}_{CS} = 9.512 \text{ psia} \tag{2.21}$$

The uncertainties of the average experimental results are now estimated using the following steps:

1. Calculate the partial derivatives of the results with respect to the independent parameters. In this experiment, the equations contain only measured values and are, therefore, in terms of the independent parameters.

$$\text{Delta Pressure, } \Delta P = P_1 - P_2 \text{ ; } \partial \Delta P / \partial P_1 = 1; \ \partial \Delta P / \partial P_2 = -1 \tag{2.22}$$

$$\text{Average Pressure, } P_{CS} = (P_1 + P_2 + P_3)/3 \text{ ; } \partial P_{CS}/\partial P_1 = \partial P_{CS}/\partial P_2 = \partial P_{CS}/\partial P_3 = 1/3 \tag{2.23}$$

2. Estimate the systematic, random, and correlated systematic uncertainties for the transducer measurements made during the experiments.

<u>Systematic Uncertainties:</u> A discussion on evaluating possible elemental systematic error sources for an experiment could become very lengthy. This topic is examined briefly in Chapter 5. However, in this application primer, the focus will be on how to combine uncertainty estimates using the equations provided to give the uncertainty of the result. Therefore, details on how to obtain the estimates of the elemental error sources will not be given here. Suppose, for the purposes of illustration, that we were able to determine the bias limits associated with each of the pressure transducers that covers all significant sources of systematic error for the experiments for each transducer. Systematic uncertainty estimates for the experimental process are as follows:

$B_{Exp_{P_1}} = 0.022$ psia, $B_{Exp_{P_2}} = 0.032$ psia, $B_{Exp_{P_3}} = 0.020$ psia, $B_{Exp_{P_4}} = 0.041$ psia, $B_{Exp_{P_5}} = 0.050$ psia

Each of the transducers was also calibrated against the same working standard prior to being used for the experiment to obtain pressure versus voltage output curves. For simplicity here, assume that the calibration resulted in a systematic uncertainty of 0.005 psia and a random uncertainty of 0.010 psia. Therefore, the uncertainty of the calibration can be calculated using Eq. 2.2 and is the same for all transducers.

$$U_{Cal} = \left(B_{cal}^2 + P_{cal}^2\right)^{1/2} = \left(0.005^2 + 0.010^2\right)^{1/2} = 0.011 \text{ psia} \tag{2.24}$$

Since the overall calibration error will be a fixed error in the experiments for each transducer, the calibration uncertainty becomes a fossilized bias error [Ref. 2.2 and Chapter 5] for the experiment. Combining the bias limits for the experimental process with the calibration uncertainty yields:

$$B_{P_1} = \left(B_{Exp\,P_1}^2 + U_{cal_{P_1}}^2\right)^{1/2} = \left(0.022^2 + 0.011^2\right)^{1/2} = 0.025 \text{ psia} \tag{2.25}$$

Similarly, $B_{P_2} = 0.034$ psia, $B_{P_3} = 0.023$ psia, $B_{P_4} = 0.042$ psia, $B_{P_5} = 0.051$ psia.

<u>Random Uncertainties:</u> The random uncertainties can be calculated directly from the multiple test results since each data point was taken over an appropriately long time interval to assume independent results. Equation 2.13 and the results in Table 2.1 produce the estimated random uncertainty for the average results obtained. The large sample assumption was used since N≥10.

$$P_{\overline{\Delta P}} = \frac{2 S_{\Delta P}}{\sqrt{N}} = \frac{0.239}{\sqrt{10}} = 0.076 \text{ psi} \tag{2.26}$$

$$P_{\overline{P}_{CS}} = \frac{2 S_{P_{CS}}}{\sqrt{N}} = \frac{0.111}{\sqrt{10}} = 0.035 \text{ psia} \tag{2.27}$$

Correlated Systematic Uncertainties: As described in the Methodology Primer, correlated systematic errors are those that are not independent of each other and account for the correlation between systematic errors in different measurements used in the calculation of a result. Correlated systematic uncertainties can occur in several ways. The correlated bias limit does not affect the uncertainty of a value measured by any one of the transducers. But, when two or more measured values are used to calculate a result, part of the bias limit in the result may be produced from the same source. In this example, assume that the uncertainty of the calibration is a common error source for all transducers, but the uncertainties for the experiment are independent and, therefore, not correlated. For these experiments, there will be correlation between the systematic errors in P_1 and P_2 for the calculation of the delta pressure and between the systematic errors in P_3, P_4, and P_5 in the calculation of the average pressure. The correlated systematic terms due to the calibration procedure are determined from Eq. 2.6 as follows:

$$B_{P_1 P_2} = (0.011)(0.011) = 0.000121 \text{ psia}^2 \tag{2.28}$$

Similarly, $B_{P_3 P_4} = B_{P_3 P_5} = B_{P_4 P_5} = (0.011)(0.011) = 0.000121 \text{ psia}^2$.

3. Calculate the bias limits, precision limits, and the uncertainties for the average of the experimental results, $\overline{P}_{CS}$ and $\overline{\Delta P}$.

Bias Limits: The bias limits for the required results are determined from Eq. 2.3 as follows:

$$B_{\Delta P}^2 = \left(\frac{\partial \Delta P}{\partial P_1} B_{P_1}\right)^2 + \left(\frac{\partial \Delta P}{\partial P_2} B_{P_2}\right)^2 + 2 \frac{\partial \Delta P}{\partial P_1} \frac{\partial \Delta P}{\partial P_2} B_{P_1 P_2} \tag{2.29}$$

$$B_{P_{CS}}^2 = \left(\frac{\partial \overline{P}}{\partial P_3} B_{P_3}\right)^2 + \left(\frac{\partial \overline{P}}{\partial P_4} B_{P_4}\right)^2 + \left(\frac{\partial \overline{P}}{\partial P_5} B_{P_5}\right)^2 + 2 \frac{\partial \Delta P}{\partial P_3} \frac{\partial \Delta P}{\partial P_4} B_{P_3 P_4} + 2 \frac{\partial \Delta P}{\partial P_3} \frac{\partial \Delta P}{\partial P_5} B_{P_3 P_5} + 2 \frac{\partial \Delta P}{\partial P_4} \frac{\partial \Delta P}{\partial P_5} B_{P_4 P_5} \tag{2.30}$$

Inputting the individual bias limits and partial derivatives and taking the square root yields,

$$B_{\Delta P} = \left((1*0.025)^2 + (-1*0.034)^2 + [2(1)(-1)(0.000121)]\right)^{\frac{1}{2}} = 0.039 \text{ psi} \tag{2.31}$$

$$B_{P_{CS}} = \left(\left(\frac{1}{3}*0.023\right)^2 + \left(\frac{1}{3}*0.042\right)^2 + \left(\frac{1}{3}*0.051\right)^2 + \left[2\left(\frac{1}{3}\right)\left(\frac{1}{3}\right)(0.000121)\right] + \left[2\left(\frac{1}{3}\right)\left(\frac{1}{3}\right)(0.000121)\right]\right.$$

$$\left. + \left[2\left(\frac{1}{3}\right)\left(\frac{1}{3}\right)(0.000121)\right]\right)^{\frac{1}{2}} = 0.025 \text{ psia} \tag{2.32}$$

The precision limits as shown in Eqs. (2.27) and (2.28) are combined with the above the bias limits to produce the estimated uncertainties in the results, $\overline{P}_{CS}$ and $\overline{\Delta P}$.

$$U_{\overline{\Delta P}} = \left((0.039)^2 + (0.076)^2\right)^{\frac{1}{2}} = 0.085 \text{ psi} \tag{2.33}$$

$$U_{\overline{P}_{CS}} = \left((0.025)^2 + (0.035)^2\right)^{\frac{1}{2}} = 0.043 \text{ psia} \tag{2.34}$$

Notice that the correlated systematic uncertainty term decreases the uncertainty of the delta pressure (the correlated systematic uncertainty term is negative) whereas the correlated systematic uncertainty terms increase the uncertainty of the average pressure (the correlated systematic uncertainty terms are positive). This example shows that the correlated systematic uncertainty terms can either increase or decrease the overall systematic uncertainty and, therefore, the overall uncertainty depends on the signs of the partial derivative terms involved.

4. Report the results of the experiment according to Sections 2.5 and 4.6 of the Ref 2.1.

The results of the experiments are:

$$\overline{\Delta P} = 5.024 \pm 0.085 \text{ psi} \tag{2.35}$$

$$\overline{P}_{CS} = 9.512 \pm 0.043 \text{ psia} \tag{2.36}$$

It should be noted that using the precision limits $\left(2S/\sqrt{10}\right)$ for the average of each pressure and propagating them through the DRE produces similar estimates for the random uncertainties calculated in Eqs. (2.26) and (2.27). This indicates that any correlated precisions that may exist are insignificant. Chapter 6 addresses this issue in more detail.

2.2.2.2 Single Test

In this example, assume that the pressure measurements for both experiments 1 and 2 were made using the same instruments as in the multiple test example. However, in this example only a single data point is measured. This is representative of the method used in many complex and costly experiments.

The experiment produced the measurements given in Table 2.2.

Table 2.2 — Measurements from Experiments 1 and 2

P_1 (psia)	P_2 (psia)	P_3 (psia)	P_4 (psia)	P_5 (psia)
12.023	6.990	9.533	9.597	9.471

$$\Delta P = P_1 - P_2 = 12.023 - 6.990 = 5.033 \text{ psi} \tag{2.37}$$

$$P_{CS} = (P_3 + P_4 + P_5)/3 = (9.533 + 9.597 + 9.471)/3 = 9.534 \text{ psia} \tag{2.38}$$

<u>Systematic Uncertainties</u>: The systematic uncertainties for the single test example are the same as for the multiple test example as shown in Eqs. (2.31) and (2.32).

<u>Random Uncertainties</u>: In this example the precision limits can be calculated from previous data. It is assumed that the test setup and instrumentation are the same as that used in the previous experiment. Using the data from Table 2.1, the large sample assumption, and Eq. (2.8) yields:

$$P_{P_1} = 2S_{P_1} = 0.154 \text{ psia} \tag{2.39}$$

Similarly, $P_{P_2} = 0.189$ psia, $P_{P_3} = 0.136$ psia, $P_{P_4} = 0.118$ psia, $P_{P_5} = 0.160$ psia.

Note that if previous data had not been available the precision limits could have been determined using any other means possible. These would include previous experience, previous data from an experiment that closely matched the current experiment, and even engineering judgment.

The precision limits for the results, ΔP and P_{CS}, are determined from Eq. 2.7 as follows:

$$P_{\Delta P}^2 = \left(\frac{\partial \Delta P}{\partial P_1} P_{P_1}\right)^2 + \left(\frac{\partial \Delta P}{\partial P_2} P_{P_2}\right)^2 \tag{2.40}$$

$$P_{CS}^2 = \left(\frac{\partial P_{CS}}{\partial P_3} P_{P_3}\right)^2 + \left(\frac{\partial P_{CS}}{\partial P_4} P_{P_4}\right)^2 + \left(\frac{\partial P_{CS}}{\partial P_5} P_{P_5}\right)^2 \tag{2.41}$$

Inputting the individual precision limits and partial derivatives and taking the square root yields,

$$P_{\Delta P} = \left((1*0.154)^2 + (-1*0.189)^2\right)^{\frac{1}{2}} = 0.244 \text{ psi} \tag{2.42}$$

$$P_{CS} = \left(\left(\frac{1}{3}*0.136\right)^2 + \left(\frac{1}{3}*0.118\right)^2 + \left(\frac{1}{3}*0.160\right)^2\right)^{\frac{1}{2}} = 0.080 \text{ psia} \qquad (2.43)$$

The bias limits as shown in Eqs. (2.31) and (2.32) are combined with the above the precision limits to produce the estimated uncertainties in the results, ΔP and P_{CS}.

$$U_{\Delta P} = \left((0.039)^2 + (0.244)^2\right)^{\frac{1}{2}} = 0.247 \text{ psi} \qquad (2.44)$$

$$U_{P_{CS}} = \left((0.025)^2 + (0.080)^2\right)^{\frac{1}{2}} = 0.084 \text{ psia} \qquad (2.45)$$

The results of the experiments are:

$$\Delta P = 5.033 \pm 0.247 \text{ psi} \qquad (2.46)$$

$$P_{CS} = 9.534 \pm 0.084 \text{ psia} \qquad (2.47)$$

Note that the uncertainties for the single test results are much larger than those for the multiple test results. The same bias limits are used for both experiments; however, the precision limits are significantly reduced for the multiple test experiment because the precision limit for an average is used. This illustrates that it is possible to reduce the uncertainty of an experimental result if multiple independent results can be obtained. Multiple test experiments also have the advantage of being able to determine the best estimate of the random uncertainty in the average results. However, it should be noted that the precision limit and uncertainty for an average result do not apply to any of the individual results. The precision limit and uncertainty for any individual result must be determined using the single test uncertainty methods.

References

2.1 "Assessment of Experimental Uncertainty with Application to Wind Tunnel Testing," AIAA Standard S-071A-1999.

2.2 Coleman, H.W. and Steele, W.G., *Experimentation and Uncertainty Analysis for Engineers, 2nd Edition*, Wiley, New York, 1999.

2.3 Brown, K.K., Coleman, H.W., Steele, W.G., and Taylor, R.P., "Evaluation of Correlated Bias Approximations in Experimental Uncertainty Analysis," *AIAA Journal*, Vol. 34, No. 5, May 1996, pp. 1013-1018.

2.4 http://www-unix.mcs.anl.gov/autodiff/ADIFOR/

2.5 Meyn, L. A., "A New Method for Integrating Uncertainty Analysis into Data Reduction Software," AIAA 98-0632, 36th Aerospace Sciences Meeting and Exhibit, Reno, NV, January 1998.

2.6 Hudson, S.T. and Coleman, H.W., "A Preliminary Assessment of Methods for Determining Turbine Efficiency," AIAA 96-0101, 34th Aerospace Sciences Meeting and Exhibit, Reno, NV, January 1996.

2.7 Markopolous, P., Coleman, H.W., and Hawk, C.W., "Uncertainty Assessment of Performance Evaluation Methods for Solar Thermal Absorber/Thruster Testing," *Journal of Propulsion and Power*, Vol. 13, No. 4, 1997, pp. 552-559.

3 General Uncertainty Analysis Example—Evaluating Measurement Methods

3.1 Introduction

A general uncertainty analysis performed in the planning phase of an experiment allows one to determine the best approach for meeting the test objectives. A general uncertainty analysis is used for the examples in both chapters 3 and 4. Often, more than one measurement method can be used to obtain the values needed in a DRE. For example, if ΔP is needed in a DRE, two absolute pressure measurements could be made, as done in the example in Chapter 2, or the ΔP could be measured directly. Since the DRE must be written in terms of the measured variables, changing the measurement method also changes the DRE. An example evaluating different measurement methods is given here in Chapter 3. Also, more than one DRE can be used to calculate the needed result. A general analysis allows one to evaluate different methods available and determine the relative importance of each measured quantity. The analysis results can then be used to choose the best DRE or measurement method and to improve critical measurements thereby reducing the uncertainty of the result. An example to evaluate different DRE's and the relative importance of each measured quantity will be shown in Chapter 4. This chapter will focus on showing all of the equations necessary to calculate the needed quantities for a general uncertainty analysis so that the process is understood. Chapter 4 will then focus on interpreting the results of a general analysis to further demonstrate its usefulness.

The following example will demonstrate how an uncertainty analysis can be used to evaluate the uncertainty associated with candidate measurement methods and the corresponding DRE's. The example shows the evaluation of three methods of determining the Mach number. Mach number is a fundamental, non-dimensional parameter for characterizing fluid flow, and its accurate determination is very important in many applications.

3.2 Mach Number Equations

The isentropic equation for calculating Mach number is:

$$M = \left[\left(\frac{2}{\gamma - 1} \right) \left[\left(\frac{P_0}{P} \right)^{\frac{\gamma-1}{\gamma}} - 1 \right] \right]^{\frac{1}{2}} \quad (3.1)$$

where,

M = Freestream Mach number

P = Freestream static pressure

P_0 = Freestream total pressure

γ = Ratio of specific heats

Therefore, to calculate Mach number, γ, P, and P_0 are needed. Here, γ is taken to have a value of 1.4, and it is assumed that the uncertainty in γ has a negligible effect on the uncertainty in Mach number relative to the effect of the uncertainties of the measured pressures.

Consider three pressure-based measurement methods that can be used to determine Mach number:

1. Absolute Based Method — measure the absolute value of both the total and static pressures.

2. Total Based Method — measure the absolute value of the total pressure and the differential pressure between the total and static pressures.

3. Static Based Method — measure the absolute value of the static pressure and the differential pressure between the total and static pressures.

The Mach number equations for the three methods then become

1. Absolute Based Method (P_0 and P are the measured variables)

$$M = \left[\frac{2}{1.4-1}\left(\left(\frac{P_0}{P}\right)^{\frac{1.4-1}{1.4}} - 1\right)\right]^{\frac{1}{2}} = 2.24\left[\left(\frac{P_0}{P}\right)^{\frac{2}{7}} - 1\right]^{\frac{1}{2}} \quad (3.2)$$

2. Total Based Method (P_0 and $\Delta P = P_0 - P$ are the measured variables)

$$M = \left[\frac{2}{1.4-1}\left(\left(\frac{P_0}{P_0 - \Delta P}\right)^{\frac{1.4-1}{1.4}} - 1\right)\right]^{\frac{1}{2}} = 2.24\left[\left(\frac{P_0}{P_0 - \Delta P}\right)^{\frac{2}{7}} - 1\right]^{\frac{1}{2}} \quad (3.3)$$

3. Static Based Method (P and $\Delta P = P_0 - P$ are the measured variables)

$$M = \left[\frac{2}{1.4-1}\left(\left(\frac{P + \Delta P}{P}\right)^{\frac{1.4-1}{1.4}} - 1\right)\right]^{\frac{1}{2}} = 2.24\left[\left(\frac{P + \Delta P}{P}\right)^{\frac{2}{7}} - 1\right]^{\frac{1}{2}} \quad (3.4)$$

With three methods available for determining Mach number, the question arises, which method yields the most accurate value of Mach number? Without performing an uncertainty analysis, it is not obvious which system will produce the lowest uncertainty for the calculated Mach number.

3.3 Uncertainty Analysis

As stated in Chapter 2, in nearly all experiments, the measured values of different variables are combined using a data reduction equation (DRE) to form some desired result (Eq. 2.1). The three Mach number equations (Eqs. 3.2-3.4) are the DRE's for this case. Each of the measured variables in the equations contain errors. These errors in the measured values then propagate through the DRE, thereby generating errors in the experimental result, Mach number. For a general uncertainty analysis, the uncertainties of the variables are not considered separately in terms of systematic and random as presented in Chapter 2. Rather, the equations are written using an overall uncertainty estimate.

Since systematic and random uncertainties are not considered separately, Eqs. 2.2, 2.3, and 2.7 are combined, and the overall uncertainty in the result, r, can be represented as

$$U_r = \left[\sum_{i=1}^{J}\left(\frac{\partial r}{\partial X_i}U_{X_i}\right)^2\right]^{1/2} \quad (3.5.1)$$

Equation 3.5.1 is equivalent to Eqs. 2.2, 2.3, and 2.7 for the general analysis where systematic and random uncertainties are not considered separately and there are no correlation terms. Dividing through by the result, r, and multiplying by X_i/X_i on the right-hand side, Eq. 3.5.1 can also be written as (Ref. 3.1)

$$\frac{U_r}{r} = \left[\sum_{i=1}^{J}\left(\frac{X_i}{r}\frac{\partial r}{\partial X_i}\right)^2\left(\frac{U_{X_i}}{X_i}\right)^2\right]^{1/2} \quad (3.5.2)$$

Here, U_{X_i} is the overall uncertainty in the variable X_i, and the interval $r \pm U_r$ contains the true (but unknown) value of r about 95 times out of 100. (Again, note that the correlation terms are neglected since systematic and random uncertainties are not considered separately.)

Applying Eq. 3.5 to Eqs. 3.2 through 3.4 yields the Mach number uncertainty equations for the three cases:

1. Absolute Based Method (P_0 and P are the measured variables)

$$U_M = \left(\left[\frac{\partial M}{\partial P_0} U_{P_0} \right]^2 + \left[\frac{\partial M}{\partial P} U_P \right]^2 \right)^{\frac{1}{2}} \quad (3.6.1)$$

$$\frac{U_M}{M} = \left(\left[\frac{P_0}{M} \frac{\partial M}{\partial P_0} \right]^2 \left[\frac{U_{P_0}}{P_0} \right]^2 + \left[\frac{P}{M} \frac{\partial M}{\partial P} \right]^2 \left[\frac{U_P}{P} \right]^2 \right)^{\frac{1}{2}} \quad (3.6.2)$$

2. Total Based Method (P_0 and $\Delta P = P_0 - P$ are the measured variables)

$$U_M = \left(\left[\frac{\partial M}{\partial P_0} U_{P_0} \right]^2 + \left[\frac{\partial M}{\partial \Delta P} U_{\Delta P} \right]^2 \right)^{\frac{1}{2}} \quad (3.7.1)$$

$$\frac{U_M}{M} = \left(\left[\frac{P_0}{M} \frac{\partial M}{\partial P_0} \right]^2 \left[\frac{U_{P_0}}{P_0} \right]^2 + \left[\frac{\Delta P}{M} \frac{\partial M}{\partial \Delta P} \right]^2 \left[\frac{U_{\Delta P}}{\Delta P} \right]^2 \right)^{\frac{1}{2}} \quad (3.7.2)$$

3. Static Based Method (P and $\Delta P = P_0 - P$ are the measured variables)

$$U_M = \left(\left[\frac{\partial M}{\partial P} U_P \right]^2 + \left[\frac{\partial M}{\partial \Delta P} U_{\Delta P} \right]^2 \right)^{\frac{1}{2}} \quad (3.8.1)$$

$$\frac{U_M}{M} = \left(\left[\frac{P}{M} \frac{\partial M}{\partial P} \right]^2 \left[\frac{U_P}{P} \right]^2 + \left[\frac{\Delta P}{M} \frac{\partial M}{\partial \Delta P} \right]^2 \left[\frac{U_{\Delta P}}{\Delta P} \right]^2 \right)^{\frac{1}{2}} \quad (3.8.2)$$

Therefore, according to Eqs. 3.6 through 3.8, the partial derivative terms and the measurement transducer uncertainties must be determined to estimate the uncertainty in the Mach number for the three methods.

3.3.1 Partial Derivative Terms

The partial derivative terms must be determined with respect to the independent parameters (measured variables). MathCad® was used to find the partial derivatives in this example, and one form of the equations is given below.

1. Absolute Based Method (P_0 and P are the measured variables)

$$\frac{\partial M}{\partial P_0} = \frac{1 + 0.2 M^2}{1.4 \, M P_0} \quad (3.9)$$

$$\frac{\partial M}{\partial P} = \frac{-\left(1 + 0.2 M^2\right)}{1.4 \, M P} \quad (3.10)$$

2. Total Based Method (P_0 and $\Delta P = P_0 - P$ are the measured variables)

$$\frac{\partial M}{\partial P_0} = \frac{-\Delta P(1+0.2M^2)}{1.4\,MP_0(P_0-\Delta P)} \tag{3.11}$$

$$\frac{\partial M}{\partial \Delta P} = \frac{1+0.2M^2}{1.4\,M(P_0-\Delta P)} \tag{3.12}$$

3. Static Based Method (P and $\Delta P = P_0 - P$ are the measured variables)

$$\frac{\partial M}{\partial P} = \frac{-\Delta P(1+0.2M^2)}{1.4\,MP(P+\Delta P)} \tag{3.13}$$

$$\frac{\partial M}{\partial \Delta P} = \frac{1+0.2M^2}{1.4\,M(P+\Delta P)} \tag{3.14}$$

3.3.2 Transducer Uncertainty

Uncertainty estimates for the pressure measurements are now needed. As stated previously, this example will use overall uncertainty estimates for the pressure measurement systems rather than separate systematic and random limits as would be used in a detailed analysis. Since we are only interested in which of the three measurement methods produces the most accurate results, the magnitude of the uncertainty of the pressure measurement system is not important. However, the same uncertainties must be used in the evaluation of each method for each type of pressure measurement system. For this example the same uncertainty estimate was used for both the absolute and differential pressure measurements.

The uncertainty of a pressure measurement system is normally quoted in one of three ways:

1. Percent of Measured Value.
2. Percent of Full Scale.
3. Percent of Full Scale plus a Percent of the Measured Value.

All three of the above methods for expressing the pressure measurement uncertainty were used to produce the results for the Mach number uncertainty given in the next section.

3.3.3 Mach Number Uncertainty

The three Mach number uncertainty equations (Eqs. 3.6-3.8), the partial derivative terms (Eqs. 3.9-3.14), and the three ways of expressing the pressure measurement system uncertainty were combined to produce the results shown in Figs 3.1-3.3. A table was set up with "data" for the P_0, P, and ΔP measurements to give the Mach number range up to 2 for each measurement method. Uncertainty estimates were then made using the 3 different quote methods described above. Again, the same uncertainty estimates were used for P_0, P, and ΔP for each of the quote methods. The maximum uncertainty for each uncertainty quote method was used to normalize the data so that the y-axis always ranges from 0 to 1. This normalization makes the magnitudes of the pressure values and uncertainty estimates irrelevant aiding the comparisons.

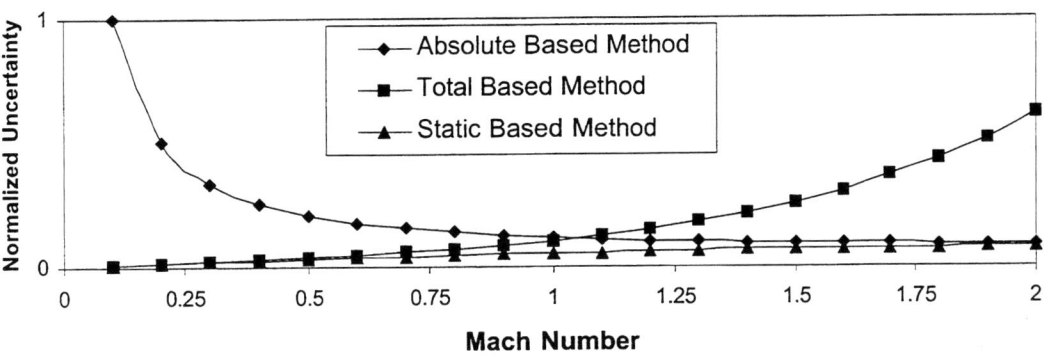

Figure 3.1 — Uncertainty in Mach Number using Pressure Uncertainty in Percent of Value

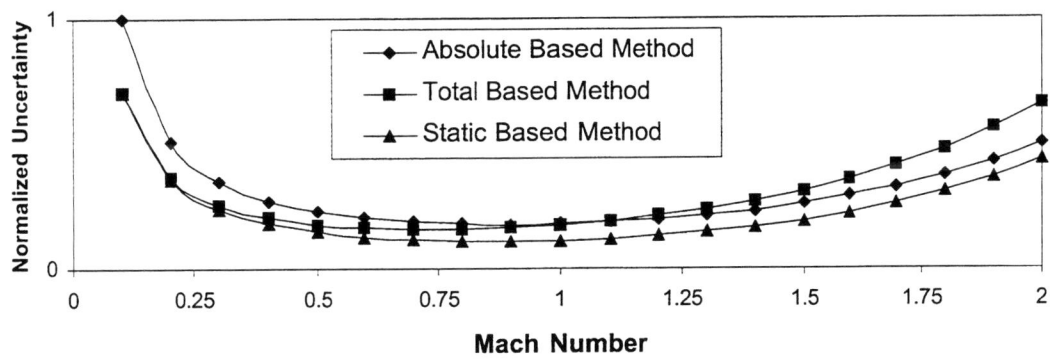

Figure 3.2 — Uncertainty in Mach Number using Pressure Uncertainty in Percent of Full Scale

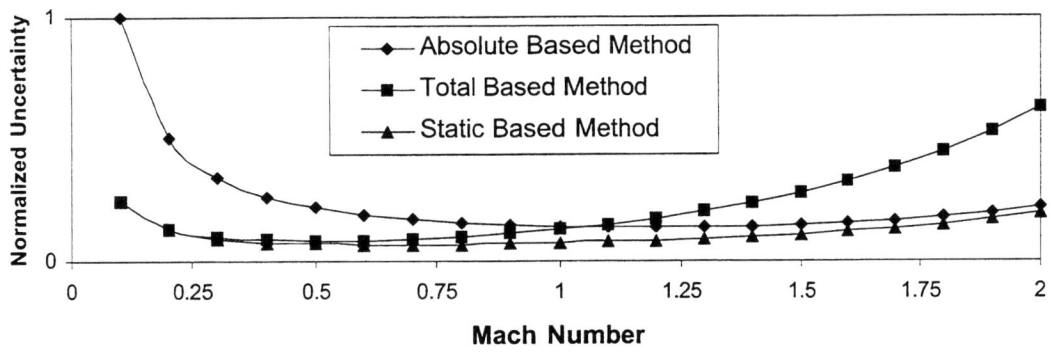

Figure 3.3 — Uncertainty in Mach Number using Pressure Uncertainty in Percent of Full Scale Plus a Percent of Value

As shown in the figures, the Static Based Method produces the lowest uncertainty for the calculation of the Mach number regardless of the way in which the pressure system uncertainty is expressed. There are Mach number ranges where either the Absolute or Total Based Methods yield the same or close to the same uncertainty, but over the entire Mach number range neither of these methods can compare to the Static Based Method. Based on the results and assumptions in this example, the **Static Based Method** produces the lowest Mach number uncertainty. It is important to remember that the uncertainty results change with different DRE's and measurement techniques.

3.3.4 Relative Sensitivity Factors

Relative sensitivity factors, described in section 2.1.3, can be used to provide further insight (Ref. 3.1, 3.2, 3.3). The UMF and UPC equations for this example will be given here simply to demonstrate how these terms are calculated. Full results and interpretation of the results for sensitivity factors will then be given in chapter 4.

The UMF terms for this case are defined by Eq. 2.15. The UMF values for the absolute based method are obtained using Eq. 2.15 along with Eqs. 3.9 and 3.10:

$$UMF_{P_0} = \frac{P_0}{M}\frac{\partial M}{\partial P_0} = \frac{P_0}{M}\left[\frac{1+0.2M^2}{1.4MP_0}\right] = \frac{1+0.2M^2}{1.4M^2} \tag{3.15}$$

$$UMF_P = \frac{P}{M}\frac{\partial M}{\partial P} = \frac{P}{M}\left[-\frac{(1+0.2M^2)}{1.4MP}\right] = -\frac{1+0.2M^2}{1.4M^2} \tag{3.16}$$

Similarly, the UMF values for the total based method are

$$UMF_{P_0} = \frac{P_0}{M}\frac{\partial M}{\partial P_0} = \frac{P_0}{M}\left[\frac{-\Delta P(1+0.2M^2)}{1.4MP_0(P_0-\Delta P)}\right] = \frac{-\Delta P(1+0.2M^2)}{1.4M^2(P_0-\Delta P)} \tag{3.17}$$

$$UMF_{\Delta P} = \frac{\Delta P}{M}\frac{\partial M}{\partial \Delta P} = \frac{\Delta P}{M}\left[\frac{(1+0.2M^2)}{1.4M(P_0-\Delta P)}\right] \tag{3.18}$$

And the UMF values for the static based method are

$$UMF_P = \frac{P}{M}\frac{\partial M}{\partial P} = \frac{P}{M}\left[\frac{-\Delta P(1+0.2M^2)}{1.4MP(P+\Delta P)}\right] = \frac{-\Delta P(1+0.2M^2)}{1.4M^2(P+\Delta P)} \tag{3.19}$$

$$UMF_{\Delta P} = \frac{\Delta P}{M}\frac{\partial M}{\partial \Delta P} = \frac{\Delta P}{M}\left[\frac{(1+0.2M^2)}{1.4M(P+\Delta P)}\right] \tag{3.20}$$

These UMF values will be the same regardless of the uncertainty quote methods since the uncertainty values are not considered in the UMF equations.

The UPC calculations now consider the uncertainty estimates. The UPC for this case is defined as in Eq. 2.16. Since we are looking at overall uncertainty and not considering the correlated terms, Eq. 2.17 is not needed. For the absolute based method, the UPC equations are

$$UPC_{P_0} = \frac{\left(\frac{P_0}{M}\frac{\partial M}{\partial P_0}\right)^2\left(\frac{U_{P_0}}{P_0}\right)^2}{\left(\frac{U_M}{M}\right)^2}*100 = \frac{(UMF_{P_0})^2\left(\frac{U_{P_0}}{P_0}\right)^2}{\left(\frac{U_M}{M}\right)^2}*100 \tag{3.21}$$

$$UPC_P = \frac{\left(\frac{P}{M}\frac{\partial M}{\partial P}\right)^2\left(\frac{U_P}{P}\right)^2}{\left(\frac{U_M}{M}\right)^2}*100 = \frac{(UMF_P)^2\left(\frac{U_P}{P}\right)^2}{\left(\frac{U_M}{M}\right)^2}*100 \qquad (3.22)$$

Similar equations can be written for the total based and static based methods. For the total based method, the equations are

$$UPC_{P_0} = \frac{\left(\frac{P_0}{M}\frac{\partial M}{\partial P_0}\right)^2\left(\frac{U_{P_0}}{P_0}\right)^2}{\left(\frac{U_M}{M}\right)^2}*100 = \frac{(UMF_{P_0})^2\left(\frac{U_{P_0}}{P_0}\right)^2}{\left(\frac{U_M}{M}\right)^2}*100 \qquad (3.23)$$

$$UPC_{\Delta P} = \frac{\left(\frac{\Delta P}{M}\frac{\partial M}{\partial \Delta P}\right)^2\left(\frac{U_{\Delta P}}{\Delta P}\right)^2}{\left(\frac{U_M}{M}\right)^2}*100 = \frac{(UMF_{\Delta P})^2\left(\frac{U_{\Delta P}}{\Delta P}\right)^2}{\left(\frac{U_M}{M}\right)^2}*100 \qquad (3.24)$$

For the static based method, the equations are

$$UPC_{P_0} = \frac{\left(\frac{P_0}{M}\frac{\partial M}{\partial P_0}\right)^2\left(\frac{U_{P_0}}{P_0}\right)^2}{\left(\frac{U_M}{M}\right)^2}*100 = \frac{(UMF_{P_0})^2\left(\frac{U_{P_0}}{P_0}\right)^2}{\left(\frac{U_M}{M}\right)^2}*100 \qquad (3.25)$$

$$UPC_{\Delta P} = \frac{\left(\frac{\Delta P}{M}\frac{\partial M}{\partial \Delta P}\right)^2\left(\frac{U_{\Delta P}}{\Delta P}\right)^2}{\left(\frac{U_M}{M}\right)^2}*100 = \frac{(UMF_{\Delta P})^2\left(\frac{U_{\Delta P}}{\Delta P}\right)^2}{\left(\frac{U_M}{M}\right)^2}*100 \qquad (3.26)$$

All terms needed to calculate the UPC values from Equations 3.21 through 3.26 have been defined previously. The UMF terms are defined in Equations 3.15 through 3.20 and the partial derivative terms are defined in equations 3.9 through 3.14. Notice that only pressure and Mach number values are needed to calculate the UMF terms. Once the UMF values have been calculated, only the uncertainty estimates for the pressure measurements and the corresponding Mach number uncertainties are needed to obtain the UPC values. UMF and UPC results will not be given and interpreted here since that is the focus of Chapter 4.

References

3.1 Coleman, H.W. and Steele, W.G., *Experimentation and Uncertainty Analysis for Engineers, 2nd Edition*, Wiley, New York, 1999.

3.2 Hudson, S.T. and Coleman, H.W., "A Preliminary Assessment of Methods for Determining Turbine Efficiency," AIAA 96-0101, 34th Aerospace Sciences Meeting, January 15-18, 1996, Reno, NV.

3.3 Markopolous, P., Coleman, H.W., and Hawk, C.W., "Uncertainty Assessment of Performance Evaluation Methods for Solar Thermal Absorber/Thruster Testing," *Journal of Propulsion and Power*, Vol. 13, No. 4, 1997, pp. 552-559.

4 General Uncertainty Analysis Example—Evaluating Data Reduction Equations

As stated in Chapter 3, a general uncertainty analysis performed in the planning phase of an experiment allows one to determine the best approach for meeting the test objectives. The details of the equations for a general uncertainty analysis were given in the previous chapter. This chapter will now focus on interpretation of the general analysis results to show the usefulness of the analysis.

An example of an uncertainty analysis done in the planning phase of an experiment that shows the benefits of performing such an analysis early in a test program is given here. (More details on this example are available in Ref. 4.1 and 4.2). Two common methods of determining efficiency (η) for cold airflow testing of a turbine were evaluated from an uncertainty analysis viewpoint. The objectives of the study were, first, to estimate the uncertainty of the efficiency (result) and, second, to determine the relative importance of the uncertainty of each of the measured quantities (variables) on the uncertainty of the efficiency. Two approaches were used to evaluate the influence of the uncertainties of each of the variables on the uncertainty of the efficiency: the **uncertainty magnification factor (UMF)** and the **uncertainty percentage contribution (UPC)** (Ref. 4.1, 4.3, 4.4). As stated in Chapters 2 and 3, the UMF approach applies to the general case and shows the effect of the uncertainty of one variable as it propagates into the result (turbine efficiency in this example). This analysis can be done very early in a program before details of the testing are known. The analysis results can be used to begin evaluating measurement methods and data reduction equations (DRE's) and to focus efforts on measurements that will have a large impact on the uncertainty of the test result. The UPC approach then incorporates the uncertainty estimates associated with a particular test situation; therefore, this type of analysis is done later in the planning phase or early in the design phase of the experiment. The analysis results can be used to further evaluate the types of measurements and DRE's for the particular experiment and to further improve the measurements that have become critical in the experiment. The UMF and UPC equations were demonstrated in Chapter 3. Now the equations will be used and the results interpreted to demonstrate their usefulness.

The general uncertainty analysis example given here was done in the planning phase of a turbine technology program to help define a test approach that would allow a test goal of $U\eta/\eta \leq 1\%$ to be met. The general analysis results were used to improve critical measurements thereby reducing the uncertainty of the efficiency. General uncertainty analysis results are presented showing a comparison between the two methods of determining turbine efficiency and the relative importance of the uncertainty of each of the measured variables on the uncertainty of the efficiency. (Note that both the UMF and UPC analyses are presented simultaneously in this example, however, in reality, the UMF analysis would generally precede the UPC analysis.)

Nomenclature

C_p Specific heat at constant pressure (BTU/lb_m°R)
h   Enthalpy (BTU/lb_m)
J   Conversion constant (778.3 ft-lb_f/BTU) or counter for measured variables in DRE
K   Conversion constant ($\pi/30$ rad*min/rev*sec)
N   Speed (RPM)
P   Pressure (psia)
Pr   Pressure ratio (total-to-total)
r   Result
T   Temperature (°R)
Tq   Torque (ft-lb_f)
U   Uncertainty estimate
$\dot{W}$ Mass flow rate (lb_m/sec)
X   Variable
γ Ratio of specific heats

η   Efficiency

Subscripts
0   Total
1   Turbine inlet
2   Turbine exit
th   Thermodynamic method
me   Mechanical method

4.1 Turbine Efficiency Equations

The two equations used to calculate turbine efficiency from measured test variables are given below. Both equations are derived from the basic definition of turbine efficiency: actual enthalpy change over ideal or isentropic enthalpy change. Both methods are used for "cold" airflow turbine testing where the temperature is relatively low so that an ideal gas may be assumed and γ and C_p are considered constant.

For the first method, the thermodynamic method, the temperature drop across the turbine is measured to determine the actual enthalpy change ($\Delta h = C_p \Delta T$). Isentropic relations are used to write the ideal enthalpy change in terms of turbine inlet and exit total pressure rather than temperature.

$$\eta_{th} = \frac{T_{0_1} - T_{0_2}}{T_{0_1}\left[1 - \left(\frac{P_{0_2}}{P_{0_1}}\right)^{\frac{\gamma-1}{\gamma}}\right]} \tag{4.1}$$

For the second method, the mechanical method, the ideal enthalpy change is calculated the same as before. However, the mechanical measurements of torque and speed are used along with the measured mass flow rate to determine the actual enthalpy change.

$$\eta_{me} = \frac{K\,Tq\,N}{J\,C_p\,\dot{W}\,T_{0_1}\left[1 - \left(\frac{P_{0_2}}{P_{0_1}}\right)^{\frac{\gamma-1}{\gamma}}\right]} \tag{4.2}$$

Note that the temperatures and pressures in Eqs. 4.1 and 4.2 are average values at a cross section.

4.2 Procedure

The two turbine efficiency equations (Eqs. 4.1 and 4.2) are the DRE's for this case. Each of the measured variables in Eqs. 4.1 and (4.2) contain errors. These errors in the measured values then propagate through the DRE, thereby generating errors in the experimental result, turbine efficiency. For a general uncertainty analysis, the uncertainties of the variables are not considered separately in terms of systematic and random as presented in Chapter 2. Rather, the equations are written using an overall uncertainty estimate. Since systematic and random are not considered separately, Eqs. 2.2, 2.3, and 2.7 are combined, and the overall uncertainty in the result, r, can be represented by Eq. 3.5.

For this example, the partial derivatives were difficult to compute analytically; therefore, the partial derivatives for the two DRE's were done numerically using a jitter program as described in Ref. 4.3. For the UPC analysis, overall uncertainty estimates for each variable were needed. The uncertainty estimates were set to realistic, achievable values based on the best information available at the time. The uncertainty values were then propagated through the DRE's to obtain the uncertainty of the efficiency calculated by the two methods. A FORTRAN code was written to calculate the efficiency, the uncertainty of the efficiency, and the sensitivity of the efficiency uncertainty to each of the measured quantities (both

UMF and UPC). Again, the UMF and UPC analyses are presented simultaneously in this example. In reality, a code would generally be written for the UMF analysis first, and this code would then be modified to add the UPC analysis later in the planning of the experiment.

4.3 Results

4.3.1 Overall Uncertainty

General uncertainty analysis results are presented showing a comparison between the two efficiency methods and the relative importance of the uncertainty of each of the measured variables on the uncertainty of the efficiency. The analysis was done for the turbine test case at the aerodynamic design point (ADP) as well as over a broad off-design operating envelope. The set point parameters for the test were the turbine inlet total pressure, inlet total temperature, speed, and total-to-total pressure ratio. The range of set point conditions used is given in Table 4.1 (ADP is set point 16). The pretest predicted values for turbine exit total temperature, torque, and mass flow rate were used (Ref. 4.5). These values along with the uncertainty estimates for each of the variables are also given in Table 4.1.

For the thermodynamic method of calculating efficiency (Eq. 4.1), the variables were the turbine inlet total pressure, exit total pressure, inlet total temperature, and exit total temperature. The uncertainty of γ was considered to be negligible for this study.

For the mechanical method of calculating efficiency (Eq. 4.2), the variables were the turbine inlet total pressure, exit total pressure, inlet total temperature, mass flow rate, torque, and speed. The uncertainties of γ, C_p, and the conversion constants were considered to be negligible for this study.

It should be noted that the uncertainties in γ and C_p (which are fossilized into systematic uncertainties) should be considered again once one moves to the detailed uncertainty analysis, and they may or may not turn out to be negligible for the detailed analysis. Also, while those uncertainties were considered not to be major contributors for this general analysis based on prior experience, they may or may not be negligible for other analyses.

A complete listing of results for each method is given in Tables 4.2 and 4.3. (Refer to Table 4.1 for the set point conditions.) Since pretest predicted values were used for the variables, the efficiencies calculated by the two different methods are the same for each set point. The calculated turbine efficiency and the uncertainty values associated with the efficiency are presented for the entire off-design matrix in Fig. 4.1 through 4.3. In general, the turbine efficiency increases as speed increases and decreases as the total-to-total pressure ratio increases (Fig. 4.1). The uncertainty in the turbine efficiency decreases as speed increases and as pressure ratio increases for both methods, as seen by the uncertainty values plotted in Fig. 4.2 and 4.3. The efficiency uncertainty estimates do not meet the test program uncertainty goal of ±1% for any of the set points, and the uncertainty values are higher for the thermodynamic method for each set point for this particular test case. These results show that improvements must be made in order to achieve the efficiency uncertainty goal for the test. The UMF and UPC results give the direction for making these improvements.

4.3.2 UMF

The thermodynamic method UMF results for each variable are plotted in Fig. 4.4 and 4.5. Note that the UMF approach does not differentiate between the turbine inlet and exit measurements—the UMF values for Po_1 and Po_2 are the same, and the UMF values for To_1 and To_2 are the same. Also, the UMF values for the two pressures are the same for each pressure ratio (Pr) set point (Fig. 4.4) since these pressure values do not change unless the pressure ratio changes. The influence of the pressure measurements decreases as the Pr increases (or the pressure drop across the turbine increases). The influence of the temperature measurements is highest at low speed and low pressure ratio where the temperature drop across the turbine is lowest (Fig. 4.5). As the turbine temperature drop increases, the UMF values for the temperature measurements decrease. The UMF values show that, in general, the temperature measurements are much more critical than the pressure measurements for accurately obtaining the turbine efficiency using the thermodynamic method; therefore, efforts should be focused on obtaining the best possible temperature measurements for the experiment.

The mechanical method UMF results for each variable are plotted in Fig. 4.6 and 4.7. Again, the UMF approach does not differentiate between the turbine inlet and exit pressures (Fig. 4.6). The UMF values for inlet total temperature, mass flow rate, torque, and speed are always one (Fig. 4.7). As with the thermodynamic method, the UMF values for the two pressures are the same for each pressure ratio set point, and these values decrease as the pressure ratio increases. Note that the UMF values for the pressure measurements are always greater than one. This data shows that, in general, the pressure measurements are critical for accurately obtaining the turbine efficiency using the mechanical method; therefore, efforts should be focused on obtaining the best possible pressure measurements for the experiment.

Suppose that, at this point in the planning of the experiment, a decision to go with one DRE or the other must be made due to time constraints for planning. If that decision needed to be made here, one would need to evaluate whether it would be easier, less costly, less time consuming, etc., to obtain accurate temperature measurements or pressure measurements for this situation. If one has more confidence in obtaining temperature measurements with low uncertainty estimates, then the thermodynamic method would be chosen. Alternatively, if one has more confidence in obtaining pressure measurements with low uncertainty estimates, then the mechanical method would be chosen. If possible, however, this decision should be delayed until later in the planning phase after studying the UPC results. The UMF results then would simply be used to study various ways of achieving low uncertainties for the critical measurements. These studies would in turn be used to estimate the uncertainty limits for the UPC analysis.

4.3.3 UPC

The thermodynamic method UPC results for each variable are plotted in Fig. 4.8 through 4.11. Again, the temperature measurements have more influence than the pressure measurements. The influence of the temperature measurements decreases slightly and the influence of the pressure measurements increases slightly as speed increases since the turbine temperature drop is increasing. This analysis shows that, for the particular situation presented here, the accuracy of the temperature measurements is critical with the exit temperature being the most important. It also shows that the accuracy of the exit pressure is slightly more significant than that of the inlet pressure. This analysis allows one to differentiate between the inlet and exit measurements whereas the UMF approach does not.

The mechanical method UPC results for each variable are plotted in Fig. 4.12 through 4.17. The influence of T_{01} is always very small, and the influence of speed is only important for N=1000 RPM. The influence of torque is small except for one extreme off-design point (N=5000 RPM and Pr=1.20) where the expected torque value is very small. The most important measurements are inlet total pressure, exit total pressure, and mass flow rate. The influence of P_{01} is important for Pr=1.20, but its significance decreases rapidly as Pr increases. The influence of P_{02} also decreases as Pr increases, but this variable remains important. For this test case, the mass flow rate measurement is the critical measurement for the design point and a large portion of the off-design range (except extreme off-design points of very low speed (1000 RPM) and low pressure ratio (1.20)). The uncertainty estimate used for the mass flow rate was 1% of the measured value. This uncertainty estimate is reasonable, but can be difficult to achieve. Therefore, the importance of the mass flow rate should be noted. The significance of the mass flow rate measurement was not apparent from the UMF analysis for a general case. Since the UMF analysis showed that the pressure measurements were important, these measurements were studied to find ways to reduce their uncertainty limits. As the uncertainties of the pressure measurements decrease, the uncertainties of the other variables potentially have more impact on the uncertainty of the result. Therefore, the uncertainty of the mass flow rate measurement became important for this particular case.

Also note the importance of the turbine exit quantities for both methods from the UPC analysis. Since the temperatures and pressures in Eqs 4.1 and 4.2 are average values at the turbine inlet and exit planes, if gradients exist in the turbine exit flowfield, then one must be particularly careful of the number of exit measurements made, the averaging techniques used, etc., in obtaining P_{02} and T_{02}. These factors can increase the uncertainty of P_{02} and T_{02} greatly (Ref. 4.2 and 4.6). Again, the UMF approach could not differentiate between the turbine inlet and exit measurements.

4.3.4 Summary

In summary, the UMF results showed that the temperature measurements were critical for the thermodynamic efficiency and the pressure measurements were critical for the mechanical efficiency. The UPC results for the thermodynamic method again showed that the temperature measurements were critical. These results further differentiated between the turbine inlet and exit measurements revealing that the turbine exit measurements were more important than the turbine inlet measurements for both pressure and temperature. The UPC results for the mechanical method again showed the importance of the pressure measurements. The exit pressure measurement was shown to have more influence than the inlet pressure measurement, as with the thermodynamic method. These results further revealed that the mass flow rate measurement was the most critical, except for extreme off-design points, for this particular test case. It should be noted that the UPC for each of the variables could change for other test situations as the variable uncertainty estimates change. This demonstrates the importance of extending the UMF study to the UPC study for each particular test situation.

For this case study, the uncertainty in turbine efficiency was always greater for the thermodynamic method. As with the UPC results, this could change for other cases when the uncertainty estimates of the variables change. It is also important to note that, since this was a general uncertainty analysis, the influence of correlated systematic uncertainties was not considered. Chapter 2 showed that the correlation effects could decrease the uncertainty for differences in measured variables (e.g., ΔP or ΔT). Since the thermodynamic method of determining efficiency involves a temperature difference, there is a potential of improving the uncertainty with this method by forcing correlation between the inlet and exit temperature measurements. A detailed uncertainty analysis must be employed to see this effect. This idea will be discussed further in Chapter 5.

The results presented here demonstrate the usefulness of both the UMF and UPC approaches. These studies should be done in the planning phase of an experiment. The UMF results allow one to focus attention on the measurements that will generally be critical to obtaining an accurate result early in the planning of an experiment. The studies done based on the UMF results then provide input for the UPC analysis. For this case, the UMF results led to studies for improving the pressure and temperature measurements in general. Then the UPC results allowed steps to be taken to improve the critical measurements of mass flow rate and temperature before the test was conducted so that the overall uncertainty in the efficiency could be improved. The UPC results also showed the importance of the turbine exit quantities; therefore, emphasis was placed on obtaining all of the measurements needed to properly mass average the data in the high gradient flowfield expected at the exit of this turbine.

References

4.1 Hudson, S.T. and Coleman, H.W., "A Preliminary Assessment of Methods for Determining Turbine Efficiency," AIAA 96-0101, 34th Aerospace Sciences Meeting, January 15-18, 1996, Reno, NV.

4.2 Hudson, S.T., "Improved Turbine Efficiency Test Techniques Based on Uncertainty Analysis Application," Ph.D. dissertation for the University of Alabama in Huntsville, 1998.

4.3 Coleman, H.W. and Steele, W.G., *Experimentation and Uncertainty Analysis for Engineers*," 2nd Edition, John Wiley & Sons, 1999.

4.4 Markopolous, P., Coleman, H.W., and Hawk, C.W., "Uncertainty Assessment of Performance Evaluation Methods for Solar Thermal Absorber/Thruster Testing," *Journal of Propulsion and Power*, Vol. 13, No. 4, 1997, pp. 552-559.

4.5 Huber, F.W., Johnson, P.D., and Montesdeoca, X.A., "Baseline Design of the Gas Generator Oxidizer Turbine (GGOT) and Performance Predictions for the Associated Oxidizer Technology Turbine Rig (OTTR)," United Technologies Pratt & Whitney SZL:38865.doc to Scientific Research Associates, Inc., April 19, 1993.

4.6 Hudson, S.T., Johnson, P.D., and Branick, R.E., "Performance Testing of a Highly Loaded Single Stage Oxidizer Turbine with Inlet and Exit Volute Manifolds," AIAA 95-2405, 1995.

Table 4.1 — General Analysis Set Point Conditions and Input Values

Pt	Set Point Conditions				Input from Predictions			
	P_{01} (psia)	T_{01} (°R)	N (RPM)	Pr_{t-t}	P_{02} (psia)	T_{02} (°R)	$\dot{W}$ (lb_m/s)	Tq (ft-lb_f)
1	100	560	1000	1.20	83.3	547.0	11.94	276.1
2	100	560	2000	1.20	83.3	540.5	11.14	193.5
3	100	560	3000	1.20	83.3	537.2	9.93	134.6
4	100	560	3710	1.20	83.3	536.0	9.18	106.0
5	100	560	4000	1.20	83.3	535.7	8.92	96.7
6	100	560	5000	1.20	83.3	535.1	8.15	72.3
7	100	560	1000	1.40	71.4	540.9	11.94	407.4
8	100	560	2000	1.40	71.4	528.3	11.94	337.2
9	100	560	3000	1.40	71.4	522.9	11.93	263.1
10	100	560	3710	1.40	71.4	519.7	11.69	226.2
11	100	560	4000	1.40	71.4	518.8	11.53	211.6
12	100	560	5000	1.40	71.4	516.5	10.92	169.3
13	100	560	1000	1.60	62.5	539.0	11.94	446.9
14	100	560	2000	1.60	62.5	523.5	11.94	388.7
15	100	560	3000	1.60	62.5	514.4	11.94	323.5
16	100	560	3710	1.60	62.5	509.3	11.94	290.7
17	100	560	4000	1.60	62.5	507.5	11.94	279.0
18	100	560	5000	1.60	62.5	503.0	11.78	239.6
19	100	560	1000	1.80	55.6	538.5	11.94	456.9
20	100	560	2000	1.80	55.6	521.4	11.94	410.5
21	100	560	3000	1.80	55.6	510.5	11.94	350.9
22	100	560	3710	1.80	55.6	503.9	11.94	321.6
23	100	560	4000	1.80	55.6	501.5	11.94	311.4
24	100	560	5000	1.80	55.6	494.4	11.85	277.2

Uncertainty Estimates (same for all set points)	
P_{01}	0.25 psia
P_{02}	0.25 psia
T_{01}	1 °R
T_{02}	1 °R
$\dot{W}$	1% of reading
Tq	1 ft-lb_f
N	15 RPM

Table 4.2 — General Analysis Thermodynamic Method Results

	Efficiency			UMF				UPC			
Pt	h	U_h	U_h %	P_{01}	P_{02}	T_{01}	T_{02}	P_{01}	P_{02}	T_{01}	T_{02}
1	0.4563	0.0501	11.0	5.3	5.3	42.2	42.2	1.5	2.1	47.1	49.3
2	0.6851	0.0509	7.4	5.3	5.3	27.8	27.8	3.2	4.6	44.4	47.7
3	0.8017	0.0515	6.4	5.3	5.3	23.6	23.6	4.3	6.2	42.9	46.6
4	0.8448	0.0518	6.1	5.3	5.3	22.3	22.3	4.7	6.8	42.3	46.1
5	0.8562	0.0519	6.1	5.3	5.3	22.0	22.0	4.9	7.0	42.1	46.0
6	0.8751	0.0520	5.9	5.3	5.3	21.5	21.5	5.1	7.3	41.8	45.8
7	0.3729	0.0275	7.4	2.8	2.8	28.3	28.3	0.9	1.8	47.0	50.3
8	0.6173	0.0278	4.5	2.8	2.8	16.7	16.7	2.5	4.8	43.7	49.0
9	0.7228	0.0281	3.9	2.8	2.8	14.1	14.1	3.3	6.5	42.0	48.2
10	0.7843	0.0282	3.6	2.8	2.8	12.9	12.9	3.9	7.6	41.0	47.6
11	0.8022	0.0283	3.5	2.8	2.8	12.6	12.6	4.0	7.9	40.7	47.4
12	0.8474	0.0284	3.4	2.8	2.8	11.9	11.9	4.4	8.7	39.9	46.9
13	0.2983	0.0199	6.7	2.0	2.0	25.7	25.7	0.6	1.4	47.2	50.9
14	0.5191	0.0201	3.9	2.0	2.0	14.3	14.3	1.7	4.2	43.9	50.2
15	0.6481	0.0202	3.1	2.0	2.0	11.3	11.3	2.5	6.5	41.6	49.4
16	0.7205	0.0204	2.8	2.0	2.0	10.0	10.0	3.1	7.9	40.3	48.7
17	0.7455	0.0204	2.7	2.0	2.0	9.7	9.7	3.3	8.4	39.8	48.5
18	0.8106	0.0206	2.5	2.0	2.0	8.8	8.8	3.8	9.8	38.5	47.8
19	0.2480	0.0161	6.5	1.6	1.6	25.1	25.1	0.4	1.2	47.3	51.1
20	0.4456	0.0162	3.6	1.6	1.6	13.5	13.5	1.2	3.7	44.2	50.9
21	0.5715	0.0163	2.9	1.6	1.6	10.3	10.3	1.9	6.1	41.8	50.3
22	0.6479	0.0164	2.5	1.6	1.6	9.0	9.0	2.4	7.7	40.2	49.7
23	0.6763	0.0164	2.4	1.6	1.6	8.6	8.6	2.6	8.4	39.6	49.4
24	0.7582	0.0166	2.2	1.6	1.6	7.5	7.5	3.2	10.3	37.9	48.6

Table 4.3 — General Analysis Mechanical Method Results

Pt	Efficiency			UMF						UPC					
	h	U_h	U_h	P_{01}	P_{02}	T_{01}	$\dot{W}$	Tq	N	P_{01}	P_{02}	T_{01}	$\dot{W}$	Tq	N
1	0.456	0.0127	2.8	5.3	5.3	1.0	1.0	1.0	1.0	22.8	32.9	0.4	12.9	1.7	29.3
2	0.685	0.0170	2.5	5.3	5.3	1.0	1.0	1.0	1.0	28.9	41.6	0.5	15.8	4.3	9.0
3	0.801	0.0200	2.5	5.3	5.3	1.0	1.0	1.0	1.0	28.7	41.4	0.5	16.3	8.9	4.1
4	0.844	0.0214	2.5	5.3	5.3	1.0	1.0	1.0	1.0	27.9	40.1	0.5	15.0	13.9	2.6
5	0.855	0.0220	2.6	5.3	5.3	1.0	1.0	1.0	1.0	27.0	38.9	0.5	15.4	16.2	2.0
6	0.874	0.0237	2.7	5.3	5.3	1.0	1.0	1.0	1.0	24.3	35.0	0.4	13.1	26.0	1.2
7	0.372	0.0082	2.2	2.8	2.8	1.0	1.0	1.0	1.0	10.4	20.4	0.7	21.0	1.3	46.1
8	0.617	0.0110	1.8	2.8	2.8	1.0	1.0	1.0	1.0	15.7	30.9	1.0	31.8	2.8	17.8
9	0.722	0.0123	1.7	2.8	2.8	1.0	1.0	1.0	1.0	17.2	33.6	1.1	34.6	5.0	8.5
10	0.783	0.0134	1.7	2.8	2.8	1.0	1.0	1.0	1.0	17.1	33.6	1.1	36.0	6.7	5.5
11	0.801	0.0138	1.7	2.8	2.8	1.0	1.0	1.0	1.0	16.9	33.2	1.1	36.6	7.5	4.7
12	0.846	0.0146	1.7	2.8	2.8	1.0	1.0	1.0	1.0	16.9	33.1	1.1	34.2	11.7	3.1
13	0.298	0.0061	2.0	2.0	2.0	1.0	1.0	1.0	1.0	5.8	14.9	0.7	23.8	1.2	53.6
14	0.518	0.0083	1.6	2.0	2.0	1.0	1.0	1.0	1.0	9.7	24.8	1.3	39.6	2.6	22.1
15	0.647	0.0098	1.5	2.0	2.0	1.0	1.0	1.0	1.0	10.9	27.8	1.4	44.4	4.2	11.4
16	0.720	0.0107	1.5	2.0	2.0	1.0	1.0	1.0	1.0	11.2	28.8	1.5	46.0	5.4	7.1
17	0.745	0.0110	1.5	2.0	2.0	1.0	1.0	1.0	1.0	11.2	28.8	1.5	46.0	5.8	6.7
18	0.809	0.0120	1.5	2.0	2.0	1.0	1.0	1.0	1.0	11.2	28.6	1.4	46.9	7.9	4.0
19	0.247	0.0050	2.0	1.6	1.6	1.0	1.0	1.0	1.0	3.8	12.3	0.8	25.1	1.2	56.9
20	0.445	0.0068	1.5	1.6	1.6	1.0	1.0	1.0	1.0	6.6	21.5	1.4	44.0	2.6	23.9
21	0.571	0.0081	1.4	1.6	1.6	1.0	1.0	1.0	1.0	7.5	24.5	1.6	50.0	4.0	12.3
22	0.647	0.0090	1.4	1.6	1.6	1.0	1.0	1.0	1.0	7.9	25.5	1.7	52.1	5.0	8.0
23	0.675	0.0094	1.4	1.6	1.6	1.0	1.0	1.0	1.0	7.9	25.5	1.7	52.2	5.3	7.4
24	0.757	0.0105	1.4	1.6	1.6	1.0	1.0	1.0	1.0	7.9	25.7	1.7	53.4	6.8	4.6

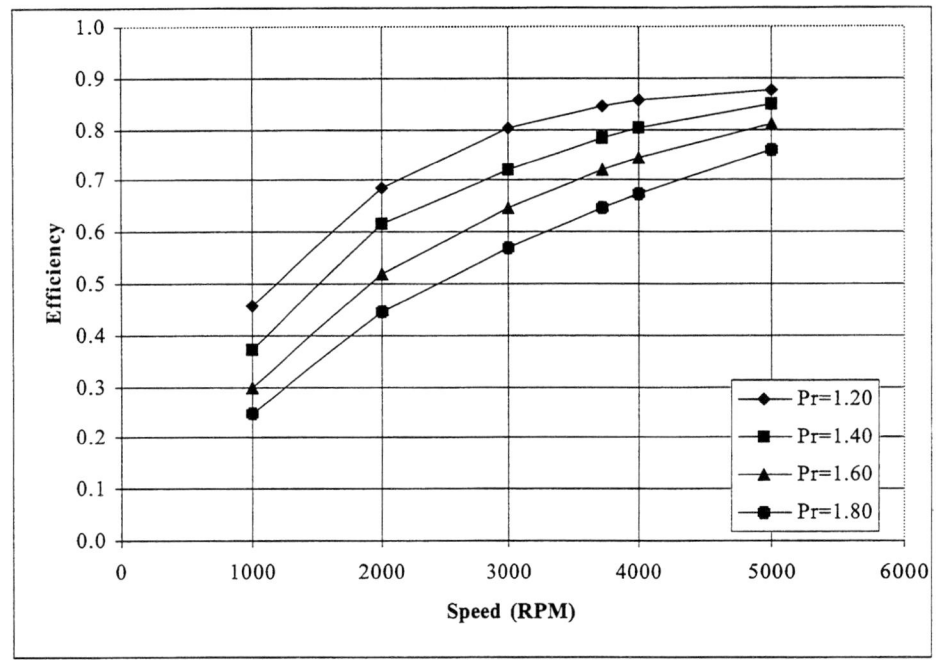

Figure 4.1 — Turbine Efficiency Over Off-Design Operating Range

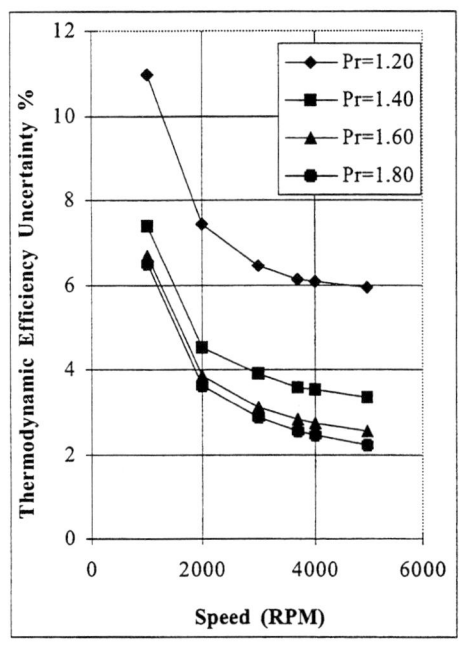

Figure 4.2 — Thermodynamic Efficiency Uncertainty (%)

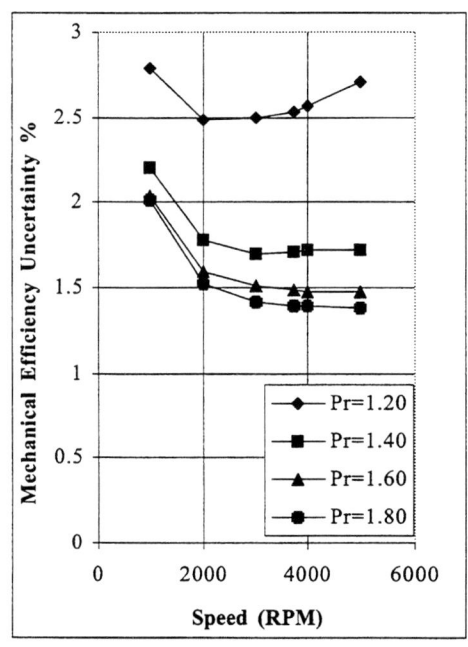

Figure 4.3 — Mechanical Efficiency Uncertainty (%)

AIAA G-045-2003

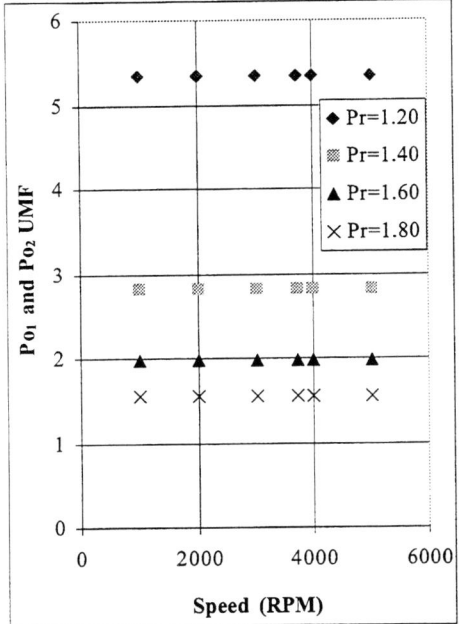

Figure 4.4 — Thermodynamic Method P_{o_1}, P_{o_2} UMF

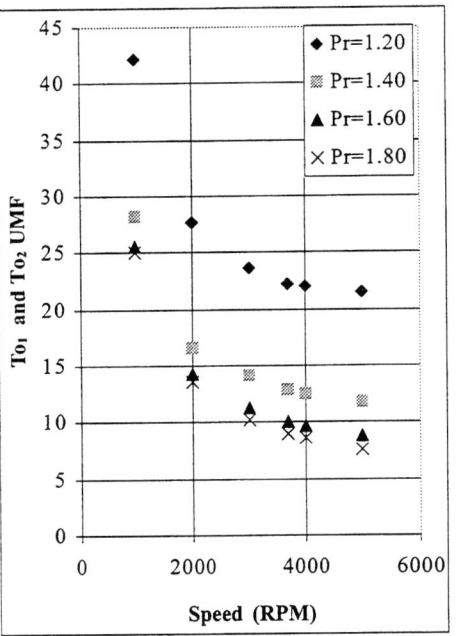

Figure 4.5 — Thermodynamic Method T_{o_1}, T_{o_2} UMF

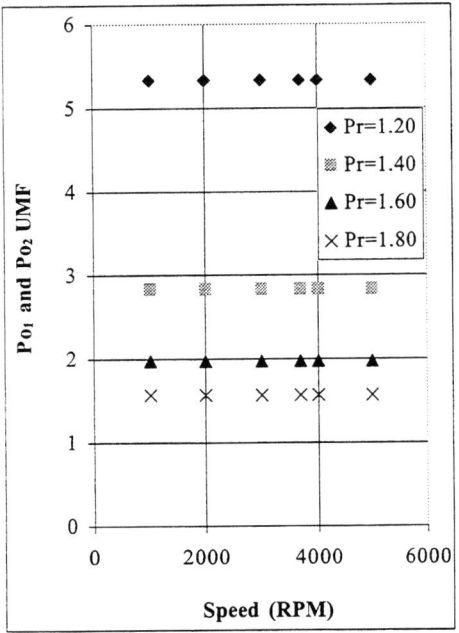

Figure 4.6 — Mechanical Method P_{o_1}, P_{o_2} UMF

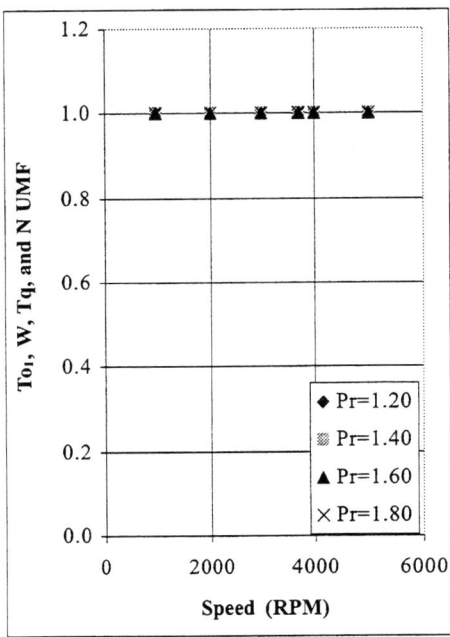

Figure 4.7 — Mechanical Method T_{o_1}, $\dot{W}$, T_q, N UMF

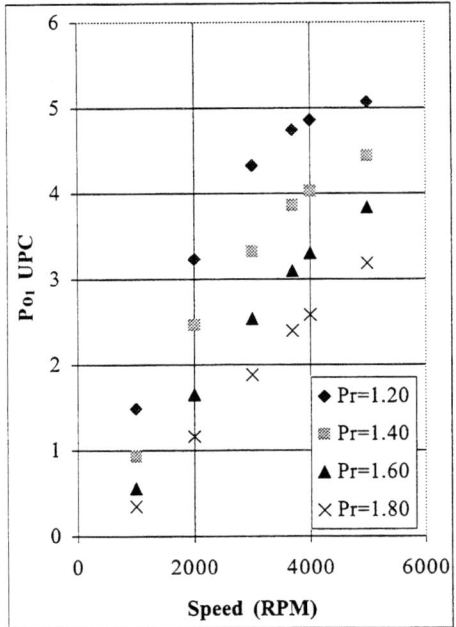

Figure 4.8 — Thermodynamic Method P_{o_1} UPC

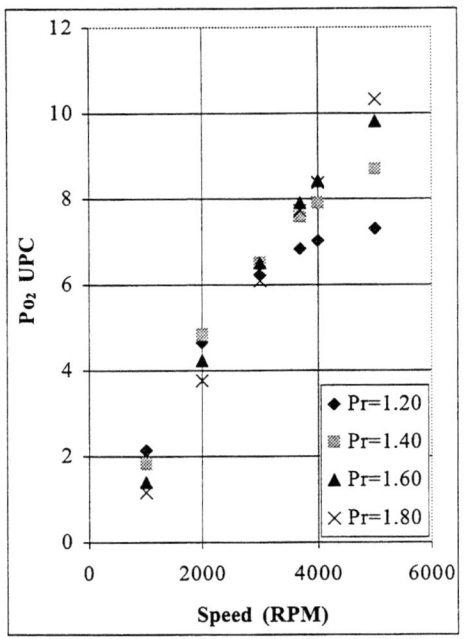

Figure 4.9 — Thermodynamic Method P_{o_2} UPC

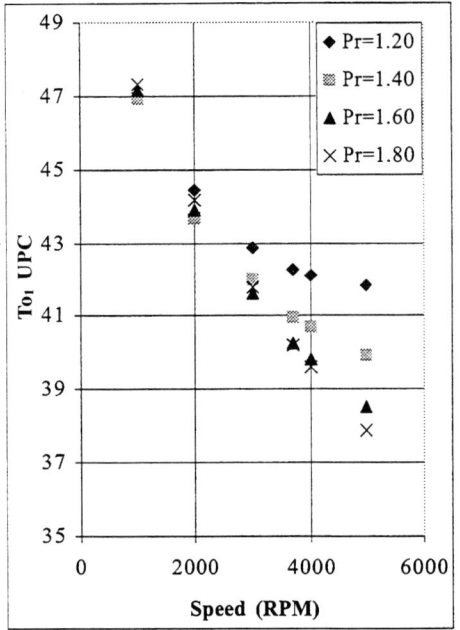

Figure 4.10 — Thermodynamic Method T_{o_1} UPC

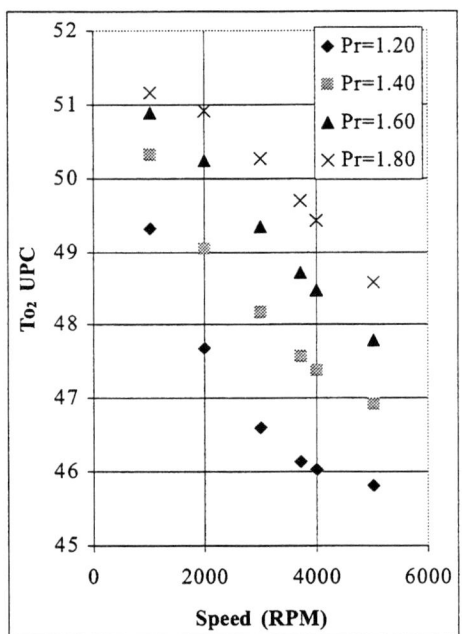

Figure 4.11 — Thermodynamic Method T_{o_2} UPC

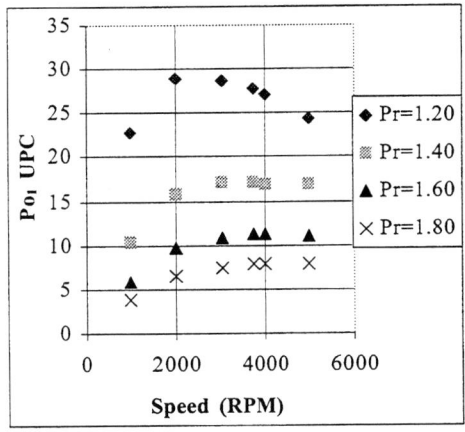

Figure 4.12 — Mechanical Method P_{o_1} UPC

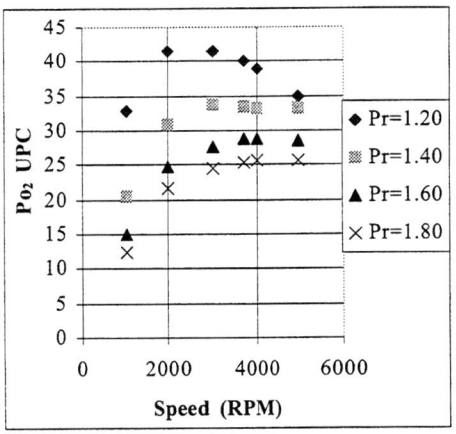

Fig. 4.13 Mechanical Method P_{o_2} UPC

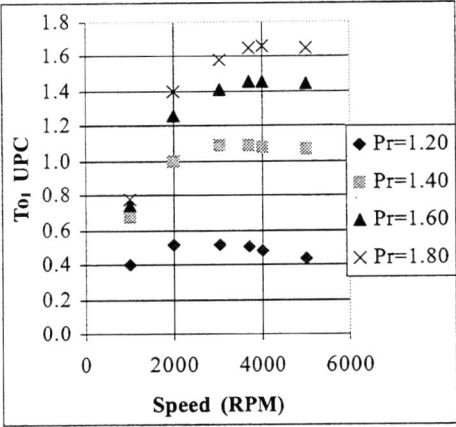

Figure 4.14 — Mechanical Method T_{o_1} UPC

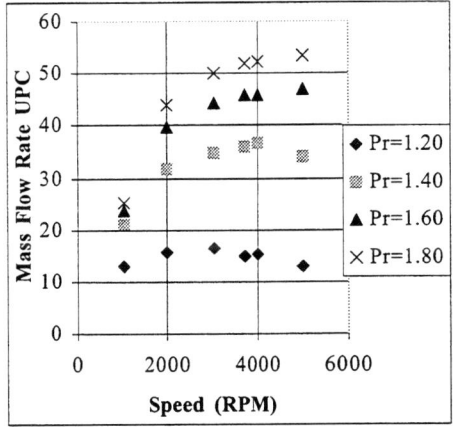

Figure 4.15 — Mechanical Method $\dot{W}$ UPC

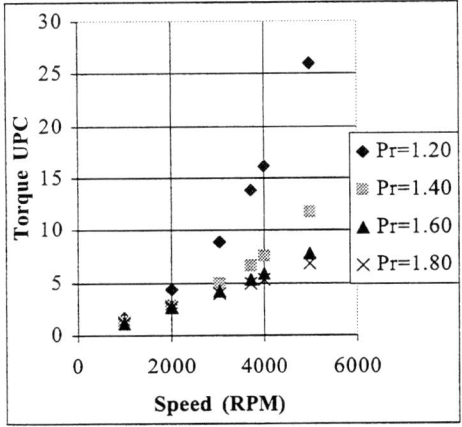

Figure 4.16 — Mechanical Method Tq UPC

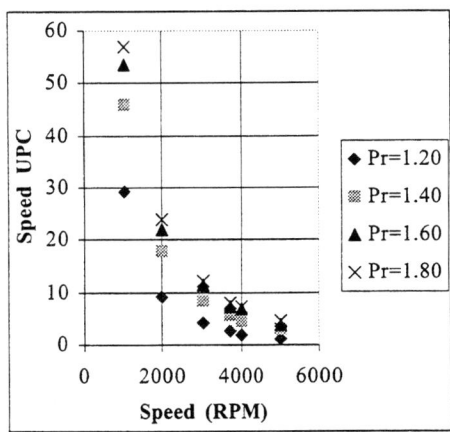

Figure 4.17 — Mechanical Method N UPC

5 Systematic Uncertainties and Correlation

This chapter now returns to a detailed uncertainty analysis, as discussed in Chapter 2, where random and systematic uncertainties are considered separately. The chapter will focus on systematic uncertainties and correlations with systematic uncertainties. The next chapter will then discuss random uncertainties and correlations with random uncertainties. Remember that correlation terms were not part of the general analyses.

This chapter begins with a brief discussion of systematic uncertainties for background information. Next, correlated systematic uncertainties are addressed. Finally, an example is given to show how correlated systematic uncertainties affect uncertainty results for various situations.

5.1 Systematic Uncertainties

To begin evaluating sources of systematic uncertainty for a particular measurement, it may be helpful to consider sources of systematic uncertainties according to these basic categories: (1) calibration errors (2) data acquisition errors (3) data reduction errors (4) conceptual errors [5.1].

(1) Calibration. A device cannot be any more accurate than the standard used for its calibration. Other uncertainty sources related to the calibration, data acquisition, etc., may then add to the uncertainty for the device. A calibration is conducted to exchange the uncertainty of the uncalibrated instrument with the uncertainty of the working standard along with the calibration process. A working standard is a device that has an uncertainty estimate that is traceable to a recognized primary laboratory, which in the U.S. is the National Institute of Standards and Technology (NIST), and should have an uncertainty that is considerably less (nominally 4 times less) than the device which is being calibrated. A calibration establishes a curve to relate the measured output of a device (e.g., volts) to the desired output (e.g., temperature). A method for estimating the uncertainty of a calibrated device is presented in Chapter 4 of the Standard [5.2].

(2) Data acquisition. The method in which the data is acquired may cause repeated readings to be in error by roughly the same amount every time. These biases may be caused by environmental and installation effects as well as biases in the data acquisition system. To demonstrate how installation effects are systematic uncertainties, consider laminar flow of air through a circular duct (Fig. 5.1). If the maximum air velocity is to be measured using a Pitot-static probe, the probe must be placed exactly in the center of the duct. Although the experimenter knows that the probe should be placed in the center of the duct, the probe will not likely be exactly in the center of the duct in practice due to uncertainties in positioning the probe. If the probe is slightly off of center, then the measurement that the probe will make will always be below the true maximum velocity of the air flow.

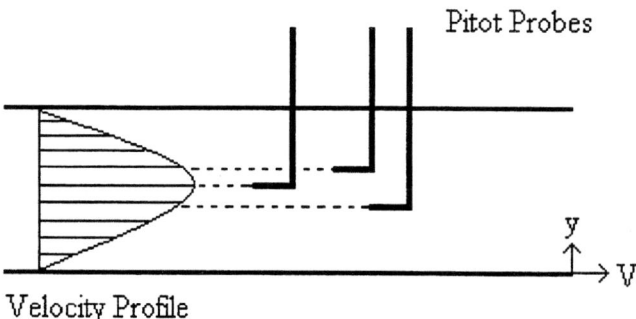

Figure 5.1 — Installation Effects

(3) Data reduction. The experimental result that is calculated from the DRE may not be the "true" result (Fig. 5.2). Often simplifying assumptions must be made to derive a DRE that models the

"real" world. The result calculated using the DRE may then differ from the "true" value due to these simplifying assumptions.

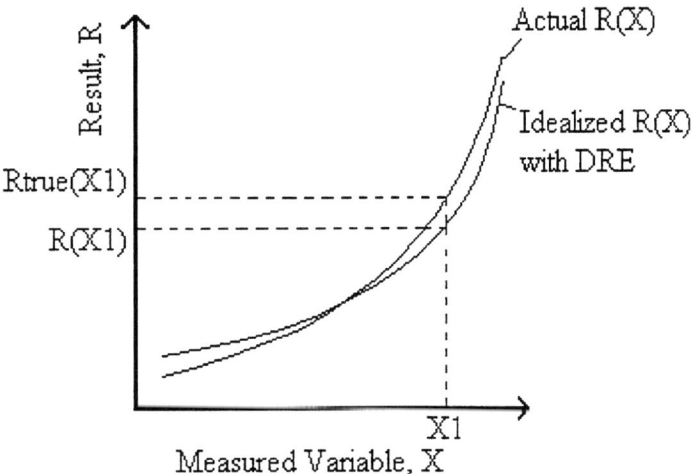

Figure 5.2 — Data Reduction Effects

Using the given DRE,

$$R_{true}(X_1) - R(X_1) = error = \text{constant} \tag{5.1}$$

for any X_1. Since the error does not change for the given X_1, the error due to the DRE is a systematic uncertainty.

(4) Conceptual Errors. These errors result when there is a difference between what is required in the DRE and what is actually measured. For example, often a cross-sectional average value is required for a DRE. However, an individual point measurement must be made or multiple point measurements must be made and averaged to obtain the value for the DRE. Refer back to Fig. 5.1, and consider that the mean velocity is required at some cross-section of the pipe. If only one total pressure measurement is made, the value of this measurement will depend on its radial position in the pipe. If the mean value is needed, the difference between the measured value and the true mean value will be a conceptual bias.

Fossilized bias errors are related to (1) above. When other laboratories use their data to determine constants, empirical relations, or tabular data, their experiments have both systematic and random uncertainties. If another experimenter uses their information (constants, empirical relation, or tabular data) in an experiment, the random uncertainties in the reported empirical information become systematic uncertainties. That is, the random uncertainties in the empirical data become "fossilized." To demonstrate this, suppose that a laboratory does some experiments to determine the relationship between the specific heat at constant pressure (c_p) and temperature for hydrogen. A curve-fit is generated from their empirical data. Using the data in this manner effectively transforms the uncertainties in the empirical information (both systematic and random) into systematic uncertainties. That is, for a given temperature (T_1), the error in the specific heat value is constant (Fig. 5.3).

$$error(T_1) = C_{p,true}(T_1) - C_{p,curvefit}(T_1) = \text{constant} \tag{5.2}$$

Since the error in the calculated specific heat value for the given temperature is constant, the error is systematic, even though the data used to make the curve-fit had both systematic and random uncertainties. The random uncertainties have become fossilized.

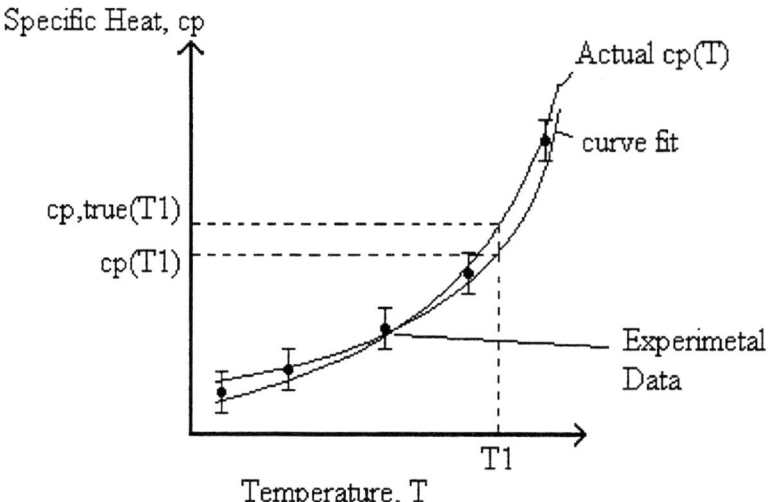

Figure 5.3 — Fossilized Bias

There are no firm rules for classifying systematic errors in the categories described above. What one experimenter considers an installation error, another experimenter may consider a conceptual error. However, that is not a problem. The usefulness of describing the above categories lies in helping the experimenter to consider all possible error sources and then to evaluate those that are significant for a particular experiment. The classification of the errors will not affect the final result since they are treated the same in the uncertainty propagation equations.

5.2   Correlated Systematic Uncertainties

Correlated uncertainties are uncertainties that are dependent on one another. The question may arise, "How do systematic uncertainties become correlated?" Two basic ways that systematic uncertainties become correlated will be covered here: (1) Measurements are made with the same instrument. (2) Two or more instruments are calibrated using the same standard at the same time or at different times.

To demonstrate how measurements made with the same instrument are correlated, consider a dial caliper. Suppose that we are about to make some length measurements for an experiment, and we want to "tare" or "zero" our instrument first. To do this, we close the jaws as far as they will go and alter the fine adjustment screw to force the caliper to read "zero" while the jaws are closed. If there unknowingly happens to be a piece of sand between the jaws during the taring process, then the distance between the jaws of the caliper is equal to the width of the piece of sand, but the length on the caliper reads "zero." If the sand is somehow removed before taking measurements, then every measurement made with the caliper will be off by the width of the grain of sand. Thus there is a systematic error in every reading taken from the caliper. Since we don't know the width of the grain of sand, we have to estimate what the systematic uncertainty will be 95% of the time. Since we cannot make measurements of anything less than one-half of the least count of our instrument, the systematic uncertainty caused by taring our instruments is usually taken to be one-half the least count. Every measurement made with the caliper will have a tare uncertainty of one-half least count that is correlated to the tare uncertainty of every other measurement made with the same caliper.

To demonstrate how calibration effects can cause correlation between measurements taken with two different instruments, consider two thermocouples. A common method of calibrating thermocouples is to immerse the thermocouple in an oil bath like the one shown in Fig. 5.4.

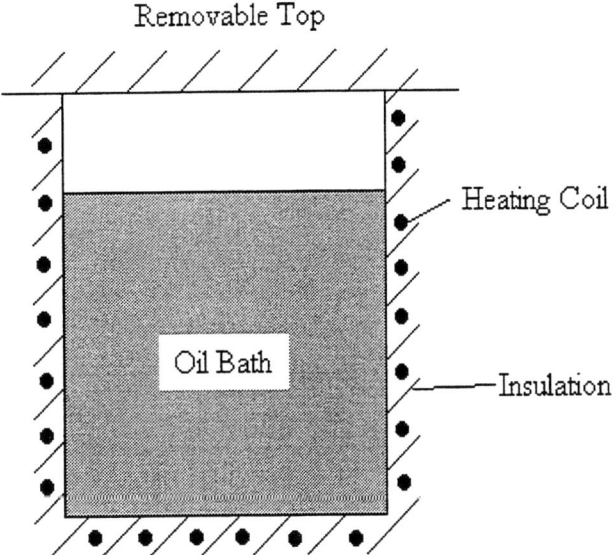

Figure 5.4 — Oil Bath for Calibration

The thermocouples are placed in the bath with a reference thermometer, and the calibration is used to force the thermocouples to read the same temperature as the reference thermometer. Because of the nature of most oils used for the bath (Prandtl number much greater than 1), it is expected that the temperature of the bath is relatively constant. However, no fluid is completely free from convective effects, and there will be some temperature variation in the fluid that is time dependent. A temperature profile of the bath at some depth from the free surface of the fluid may look something like the profile in Fig. 5.5.

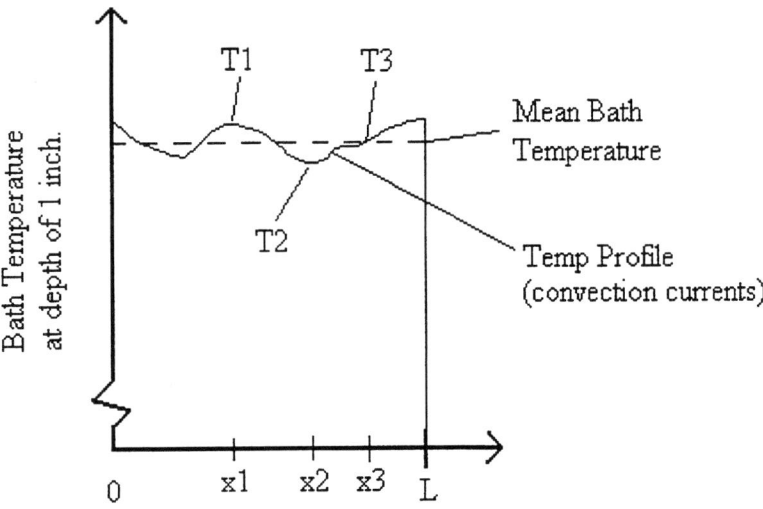

Figure 5.5 — Bath Temperature Profile

5.2.1 Bath Uniformity

As a hypothetical experiment, assume that the reference thermometer is completely accurate and, therefore, has no uncertainty. Suppose that a thermocouple to be calibrated is placed at location x1 while the reference thermocouple is placed at location x3. The actual temperature at x1 is slightly higher than at x3. If the thermocouple is calibrated to read the temperature of the reference thermometer, then all readings taken using that thermocouple will be slightly lower than the true temperature experienced by the thermocouple. The opposite occurs for a thermocouple placed at position x2 for calibration. The actual temperature at x2 is slightly lower than at x3. If the thermocouple is calibrated to read the temperature of the reference thermometer, then all readings taken using that thermocouple will be slightly higher than the true temperature experienced by the thermocouple.

Now suppose that we take two thermocouples and place them as close together as possible at position x1. (Again, assume that the reference thermometer has no uncertainty; therefore, only bath uniformity is being considered.) If we calibrate both thermocouples to read the temperature of the reference thermometer at the same time, then the systematic errors of measurements made using both thermocouples will be correlated. The errors will be equal and, because the thermocouples were calibrated at the same time, the errors are dependent on each other. The errors must be equal and dependent on one another to be correlated. If the two thermocouples were both calibrated at position x1 but at different times, then the errors may not be correlated. If the two thermocouples were calibrated at the same time but were placed in different positions in the bath (e.g., thermocouple one was at x1 while thermocouple two was at x2), then the errors will not be correlated. (Remember that only the bath uniformity is being considered here since the uncertainty of the standard was assumed to be zero.)

5.2.2 Standard Uncertainty

For this hypothetical experiment, suppose that now the temperature of the oil bath is constant but the standard has an uncertainty. Therefore, the only systematic uncertainty being considered now is that due to the reference thermometer. If the reference thermometer has a systematic error and a thermocouple is calibrated to read the value of the reference thermometer, then every time that a reading is taken with the thermocouple, it will have the same systematic error as the reference thermometer. Also, every thermocouple that is calibrated using the same reference thermometer will have the systematic uncertainty of the reference thermometer. Therefore, if two thermocouples are calibrated against the same reference thermometer, the systematic uncertainties of the thermocouples are equal and dependent on each other (because they are dependent on the reference thermometer). Then the systematic uncertainties due to the reference thermometer are correlated for the two thermocouples. (Remember that only the standard uncertainty is being considered here since the uncertainty due to the bath uniformity was assumed to be zero.)

5.3 Correlated Systematic Uncertainties Example

The pressure measurements example in Chapter 2 considered correlated systematic uncertainties. Correlation between the systematic errors was assumed between P_1 and P_2 for the ΔP calculation and between P_3, P_4, and P_5 for the P_{CS} calculation. The results showed that the correlation terms could have different effects on the systematic uncertainties of the calculated results depending on the DRE. For the upcoming example, we will see how the correlation terms are calculated when four different experimental procedures are followed for a temperature difference. It will be shown that the values of the correlation terms depend on the procedures used. The different procedures, therefore, result in different systematic uncertainty values.

Suppose we are trying to measure the change in temperature of a fluid passing through a heat exchanger using thermocouples. The temperature change is given by the following equation:

$$\Delta T = T_2 - T_1 \qquad (5.3)$$

Because of spatial variations in the temperature of the inlet and exit streams, conceptual biases of $\pm 0.1°$ C at the inlet (state 1) and $\pm 0.2°$ C at the exit (state 2) are estimated. The manufacturer of the thermocouples specifies that the thermocouples have a systematic uncertainty of $\pm 1.0°$ C.

Now, the systematic uncertainty of ΔT will be calculated for four different scenarios.

Procedure 1: Determine the uncertainty of the measurement of ΔT if the thermocouples are not recalibrated.

For this case, the heat exchanger spatial variation and the as-delivered probe spec are combined using Eq. 2.5 to obtain systematic uncertainty estimates for T_1 and T_2.

$$B_{T_1} = \left[(1.0)^2 + (0.1)^2\right]^{\frac{1}{2}} = 1.0°C \quad (5.4)$$

$$B_{T_2} = \left[(1.0)^2 + (0.2)^2\right]^{\frac{1}{2}} = 1.0°C \quad (5.5)$$

The DRE is given by Eq. (5.3). The uncertainty analysis expression [from Eq. (2.3)] is

$$B^2_{\Delta T} = \theta^2_{T_1} B^2_{T_1} + \theta^2_{T_2} B^2_{T_2} + 2\theta_{T_1}\theta_{T_2} B_{T_1 T_2} \quad (5.6)$$

where

$$\theta_{T_1} = \frac{\partial \Delta T}{\partial T_1} = -1 \quad (5.7)$$

$$\theta_{T_2} = \frac{\partial \Delta T}{\partial T_2} = 1 \quad (5.8)$$

For this case, when we do not calibrate the probes, the systematic uncertainties affecting the probes are independent of one another and

$$B_{T_1 T_2} = 0 \quad (5.9)$$

Therefore, the systematic uncertainty in ΔT is

$$B^2_{\Delta T} = (-1)^2 (1.0)^2 + (1)^2 (1.0)^2 \quad (5.10)$$

or $B_{\Delta T} = 1.4°C$.

Procedure 2: Determine the uncertainty of the measurement of ΔT if the thermocouples are recalibrated at different times with two different reference thermometers. Each reference thermometer has an accuracy of ±0.3° C in a bath with a spatial variation of ±0.1° C.

In this case, the uncertainty from the manufacturer's spec will be replaced by the calibration uncertainty. The heat exchanger spatial variation, the calibration reference thermometer uncertainty, and the calibration bath spatial variation are combined using Eq. (2.5) to obtain systematic uncertainty estimates for T_1 and T_2.

$$B_{T_1} = \left[(0.1)^2 + (0.3)^2 + (0.1)^2\right]^{\frac{1}{2}} = 0.33°C \quad (5.11)$$

$$B_{T_2} = \left[(0.2)^2 + (0.3)^2 + (0.1)^2\right]^{\frac{1}{2}} = 0.37°C \quad (5.12)$$

Again, the DRE is given by Eq. (5.3), and the uncertainty analysis expression is given by Eq. (5.6).

For this case when we calibrate the probes at different times with different reference thermometers, the systematic uncertainties affecting the probes are independent of one another and

$$B_{T_1 T_2} = 0 \quad (5.13)$$

Therefore, the systematic uncertainty in ΔT is

$$B^2_{\Delta T} = (-1)^2 (0.33)^2 + (1)^2 (0.37)^2 \quad (5.14)$$

or $B_{\Delta T} = 0.50°C$.

The systematic uncertainty of ΔT is lower for this case since the calibration uncertainty is less than the uncertainty specified by the manufacturer. The correlation terms have not had an affect in either one of these cases.

Procedure 3: Determine the uncertainty of the measurement of ΔT if the thermocouples are recalibrated at the same time using the same thermometer with an accuracy of ±0.3° C but at different locations in a bath with a spatial variation of ±0.1° C.

For this case, the systematic uncertainties in T_1 and T_2 are again given by Eqs. (5.11) and (5.12), the DRE is given by Eq. (5.3), and the uncertainty analysis expression is given by Eq. (5.6). However, now the systematic uncertainty due to the calibration reference thermometer is the same for both thermocouples. Therefore,

$$B_{T_1 T_2} = (0.3)(0.3) = 0.09 \tag{5.15}$$

Now, according to Eq. (5.6) we obtain

$$B^2_{\Delta T} = (-1)^2 (0.33)^2 + (1)^2 (0.37)^2 + (2)(-1)(1)(0.09) \tag{5.16}$$

or $B_{\Delta T} = 0.26°$ C.

In this case, the correlation term further reduced the systematic uncertainty in ΔT. Remember, as in Chapter 2, the correlation reduces the uncertainty when a difference is involved.

Procedure 4: Determine the uncertainty of the measurement of ΔT if the thermocouples are recalibrated at the same time using the same thermometer with an accuracy of ±0.3° C in a bath with a spatial variation of ±0.1° C while holding the two thermocouples in the same location in the bath.

For this case, the systematic uncertainties in T_1 and T_2 are again given by Eqs. (5.11) and (5.12), the DRE is given by Eq. (5.3), and the uncertainty analysis expression is given by Eq. (5.6). However, now the systematic uncertainties due to the calibration reference thermometer <u>and</u> the calibration spatial variation are the same for both thermocouples. Therefore,

$$B_{T_1 T_2} = (0.3)(0.3) + (0.1)(0.1) = 0.10 \tag{5.17}$$

Now, according to Eq. (5.6) we obtain

$$B^2_{\Delta T} = (-1)^2 (0.33)^2 + (1)^2 (0.37)^2 + (2)(-1)(1)(0.10) \tag{5.18}$$

or $B_{\Delta T} = 0.21°$ C.

In this case, the correlation term further reduced the systematic uncertainty in ΔT.

As shown in this example, the procedures used greatly affect the uncertainty estimates. Also, the correlation terms can have a significant impact on the uncertainty estimates. It is crucial that the uncertainty analysis calculations reflect the correct procedures for the particular situation. Also, correlation terms can be forced to occur in situations where they are helpful and avoided in situations where they are harmful with proper analysis and planning. This was done for the case reported in Chapter 4. Remember that the correlation terms are not considered in a general analysis, but the general analysis results did reveal that the temperature measurements were critical in accurately determining turbine efficiency using the thermodynamic method. Therefore, a calibration procedure was developed to force correlation between the turbine inlet and exit temperature measurements that greatly reduced the uncertainty in the thermodynamic efficiency [5.3 and 5.4]. Uncertainty analysis techniques applied in the planning phase allowed great improvements to be made in the experiment and the test goals to be met. Otherwise, experimenters would have been unaware of the problems with the data. They would not have recognized that the test goals had not been met until the data had been obtained, analyzed, and a posttest uncertainty analysis completed. Obviously, that would have been too late!

References

5.1. Coleman, H.W. and Steele, W.G., "Experimentation and Uncertainty Analysis for Engineers," 2nd Edition, John Wiley & Sons, 1999.

5.2. "Assessment of Experimental Uncertainty with Application to Wind Tunnel Testing," AIAA Standard S-071A-1999.

5.3. Hudson, S.T. and Coleman, H.W., "Analytical and Experimental Assessment of Two Methods of Determining Turbine Efficiency," *Journal of Propulsion and Power*, Volume 16, Number 5, September-October 2000.

5.4. Hudson, S.T., "Improved Turbine Efficiency Test Techniques Based on Uncertainty Analysis Application," Ph.D. dissertation for the University of Alabama in Huntsville, 1998.

Part II Advanced Topics
6 Random Uncertainties and Correlation

The remainder of the chapters in this Guide deal with more advanced topics in uncertainty analysis. In these chapters, the details of all of the calculations cannot be shown. However, enough information will be given to make the reader aware of these ideas and give them guidance in application.

This Guide began with a detailed uncertainty analysis of a simple example in Chapter 2. Correlated systematic uncertainty (bias) terms were considered in that analysis, but the correlated random uncertainty (precision) terms were zero. Chapters 3 and 4 then presented general uncertainty analysis examples where correlated terms were not involved. Chapter 5 returned to a detailed analysis that considered systematic uncertainties with correlation. It was seen that these terms can be very important in determining uncertainty. Similarly, there are cases where correlated random uncertainty terms can be very important in determining uncertainty. The example in this chapter is summarized from reference 6.1 and shows that an uncertainty analysis produced some unanticipated results that could have impacts during certain test situations. These unanticipated results were due to correlated random uncertainty terms.

The chapter will focus on random uncertainties and correlations with random uncertainties. It begins with a brief discussion of random uncertainties for background information. Then, the example is given to show how correlated random uncertainties can be important in certain situations.

6.1 Random Uncertainties

A random error is a variable error that cannot be reduced by calibration but can be reduced by multiple readings. The ways in which random uncertainties are determined vary depending on the experiment and the phase of the experiment. For example, in the planning phase of an experiment, manufacturer's information may be used to estimate the random uncertainties. When obtaining test data, repeated measurements may be used to estimate random uncertainties based on the equations given in Chapter 2. If repeated measurements cannot be made during the actual experiment, then previous information may be used to estimate the random uncertainties (e.g., from a similar previous test or calibration data). More information on determining random uncertainties is available in reference 6.2.

6.2 Correlated Random Uncertainty Example

6.2.1 Background

Mass flow venturis are used in a variety of test situations to determine the mass flow rate of gases or liquids. To accurately determine the mass flow rate using a venturi, the discharge coefficient, C_d, must be known. The discharge coefficient is defined as the ratio of the actual mass flow passing through the venturi to the ideal one-dimensional inviscid mass flow rate. Critical flow venturis are often used since C_d is well known for standard designs such as the Smith-Matz [Ref. 6.3] and the ASME [Ref. 6.3, 6.4] nozzles. However, a critical flow venturi is not always the most desirable option [Ref. 6.4]. When subsonic flow meters must be used, determination of the discharge coefficient is crucial to meter accuracy.

A detailed uncertainty analysis was preformed on venturi calibration data to determine the uncertainty of the discharge coefficient. This analysis yielded results that were inconsistent with a pretest uncertainty analysis. Random uncertainty estimates using the propagation equation (Eq. 2.7) with the covariance, P_{ik}, set to zero were an order of magnitude larger than random uncertainty estimates calculated directly from a sample of results (discharge coefficient) obtained at the same experimental set point. The differences were due to the effect of correlated random errors, which had previously been considered negligible. The significance of the correlated random uncertainties could apply to many test situations.

For this example, the discharge coefficient from the calibration needed to be determined within approximately 0.25%. The question was whether or not the venturi could be calibrated to the required accuracy. A nine-point calibration test matrix consisting of three venturi throat Mach numbers (0.2, 0.5, and 0.7) and three venturi throat Reynolds numbers (1, 3, and 6 million) was decided upon. A test group was asked to evaluate the uncertainty in C_d that they could obtain in their test facility for this test matrix. The test group estimated a systematic uncertainty and a random uncertainty for each measurement at each test point. These estimates were then propagated through the DRE for C_d to estimate an uncertainty in the discharge coefficient. Based on these estimates, it appeared that the calibration goal could be achieved at Mach numbers of 0.5 and 0.7 with very careful calibration. However, the low Mach number (0.2) points seemed to be a problem. Large random uncertainties were predicted at these points due to the low pressure differential between the venturi inlet and throat at these conditions. Based on past experience, however, the group felt that they could achieve lower random uncertainty limits with their test system than those predicted. It was decided to proceed with the calibration and to determine the random uncertainty limit in the discharge coefficient directly from the test data [Ref. 6.2]. Random uncertainty limits determined from the test data would then be combined with the systematic uncertainty limits to determine an overall uncertainty for each test point. These uncertainties would be compared with the pretest predictions and calibration requirement.

6.2.2 Calibration Approach and Results

This section will briefly describe the venturi as well as the calibration experimental setup, measurements, and test procedure. The results of the calibration will then be given.

The venturi was a Herschel-type venturi as defined in reference 6.5. The nominal inlet diameter was 7.980 inches, and the nominal throat diameter was 3.396 inches. This venturi was sized to operate at subsonic conditions to minimize flow noise or pressure fluctuations caused by a normal shock. The venturi was made of stainless steel and had two instrumentation planes. The venturi inlet plane contained two gas temperature probes and four static pressure taps. The venturi throat contained four static pressure measurements. The overall length of the venturi was 59 inches.

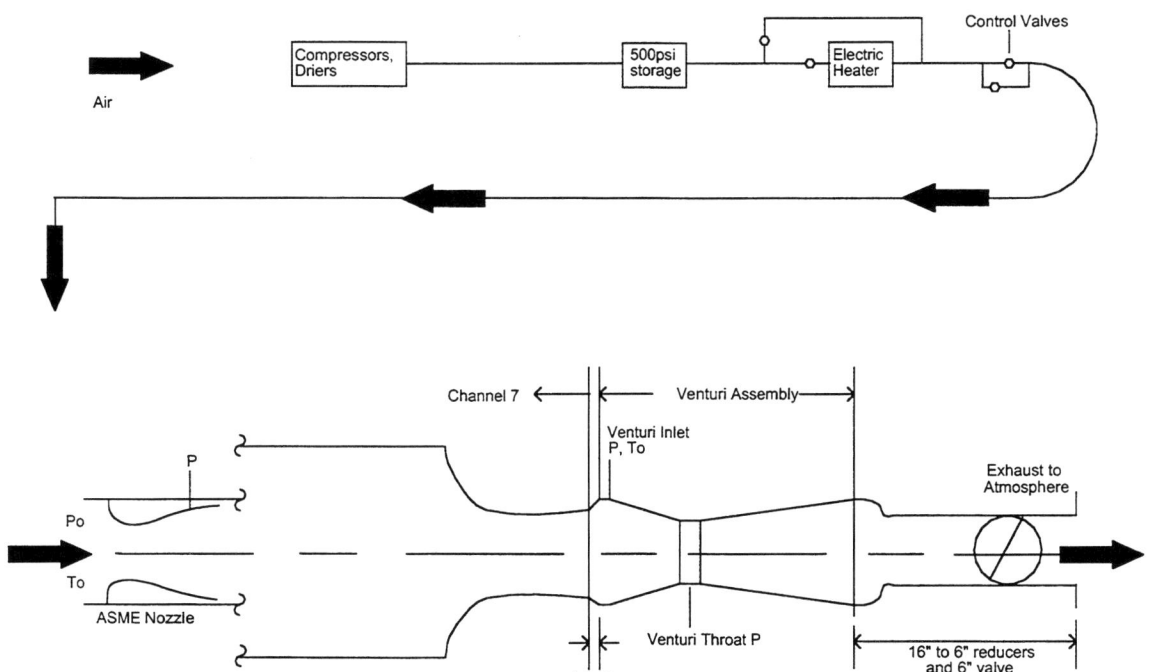

Figure 6.1 — Venturi Calibration Schematic (not to scale)

The facility used for the venturi calibration consisted of an air source, heater, inlet control valve, ASME metering nozzle, settling chamber, test venturi, and back pressure control valve. Two ASME long-radius metering nozzles were used as the calibration standards to cover the range of calibration conditions. The

ASME nozzle nominal throat diameters were 1.47 inches and 1.75 inches. Both nozzles were operated at choked conditions for all test runs (Fig. 6.1).

Pressure and gas temperature measurements were recorded for both the metering nozzle (calibration standard) and test venturi. The inlet total pressure, two inlet gas temperatures, and a static pressure just downstream of the throat were measured on the metering nozzle. For the venturi, the four inlet static pressures were averaged to obtain the inlet static pressure, the four throat static pressures were averaged to obtain the throat static pressure, and the two inlet gas temperatures were averaged to obtain the inlet temperature. Atmospheric pressure was also measured for each run. The pressure measurements were made with an electronic pressure scanning system. The metering nozzle and the venturi gas temperatures were measured with thermocouples. The voltage outputs from the thermocouples were measured using an electronic digital data acquisition system.

The venturi was calibrated over the range of throat Mach numbers and throat Reynolds numbers by setting the metering nozzle upstream pressure (mass flow for choke condition) with the upstream control valve and the test venturi back pressure with the back pressure control valve. With the metering nozzle choked, the test mass flow and venturi Reynolds number were fixed for a constant metering nozzle inlet total pressure and temperature. The venturi back-pressure was then adjusted to set the venturi Mach number. All test runs were conducted with the air at room temperature.

Once the test conditions were established, the pressures and temperatures were measured ten times (10 data scans). The time scale between measurements was such that each of the ten measurements was considered an independent reading. The venturi pressures and temperatures were recorded and used to calculate M, Re, and C_d for each set of measurements or data scan. The equations for these calculations are closed form and are given in the Appendix to this chapter [Ref. 6.5]. The ten data scans were then averaged to obtain mean values of M, Re, and C_d. Three runs were made (10 scans each, except run 3 which, for unknown reasons, had 9 scans) for the "worst case" data point (M=0.2, Re=1x10^6). This was the point where the uncertainty was expected to be the highest. These repeat runs were made to provide more data for understanding the random uncertainty behavior associated with this point. The average values of M, Re, and C_d for each run are given in Table 6.1. (Note: Due to facility limitations at M=0.7, Re had to be increased to 1.45 million to achieve the test point.)

Table 6.1 — Venturi Calibration Results

Run / # of Scans	M	Re x 10^{-6}	C_d
1 / 10	0.197	0.975	0.9899
2 / 10	0.192	0.966	0.9917
3 / 9	0.201	1.084	0.9866
4 / 10	0.199	2.932	0.9894
5 / 10	0.199	5.953	0.9933
6 / 10	0.502	0.977	0.9885
7 / 10	0.495	2.971	0.9914
8 / 10	0.493	5.834	0.9927
9 / 10	0.698	1.450	0.9907
10 / 10	0.695	2.951	0.9914
11 / 10	0.683	5.894	0.9938

6.2.3 Uncertainty Analysis

An uncertainty analysis was performed on the calibration data to determine the uncertainty of the discharge coefficient (U_{Cd}). The uncertainty analysis equations presented in Chapter 2 were used for this analysis. Traditionally, the random uncertainty correlation terms in Equation 2.7 have been assumed to be negligible. This assumption has been made since random errors are considered "random." However, the "random" assumption means that there is no correlation in sequential measurements of a single variable. But, the primary issue concerns the correlation of random errors in simultaneous measurements of different variables. This example shows that these correlation terms may be significant after all.

Some previous approaches [Ref. 6.6, 6.7, 6.8] considered the best estimate of the result to be determined using averages of the X_i so that

$$r = r(\overline{X}_1, \overline{X}_2, ..., \overline{X}_J) \tag{6.1}$$

and specified that (for large samples) the random uncertainty limits be taken as $P_i/(N_i)^{1/2}$. Equation 2.7 is then used to calculate the random uncertainty limit (with the correlation terms neglected). The P_{ik} terms in Equation 2.7 take into account the possibility of random errors in different variables being correlated, and these terms have traditionally been neglected, as mentioned previously. However, random errors in different variables caused by the same uncontrolled factor(s) are certainly possible. In such cases, one would need to acquire sufficient data to allow a valid statistical estimate of the correlation coefficients to be made in Equation 2.7.

Rather than averaging the X_i's as in Equation 6.1, if a test is performed so that m multiple sets of measurements $(X_1, X_2, ..., X_J)_k$ at the same test condition are obtained, then m results can be determined using Equation 2.1, and the best estimate of the result r would be $\overline{r}$ (Eq. 2.12). If the m sets of measurements were obtained over an appropriate time period, the random uncertainty limit that should be associated with this averaged result would be $P\overline{r}$ (Eq. 2.13). For a single result, Eq. 2.14 gives the standard deviation of the sample. Note that the approach using Eqs. 2.12 through 2.14 implicitly includes the correlated error effect — a substantial advantage over the approach of Equation 6.1.

At each test point, two methods were used to evaluate the random uncertainty limit associated with the experimental result, C_d. First, the random uncertainty limit of the result was calculated by propagating the random uncertainty limits of the averaged measured variables through the DRE while neglecting correlated random errors (method 1). Equations 2.7 through 2.11 and Equation 6.1 were used for this approach. (Note that this is the method used in the first part of the example in Chapter 2 — Section 2.2.2.1.) Second, the random uncertainty limit of the result was calculated directly from the sample standard deviation of C_d (method 2). Equations 2.12 through 2.14 and Equation 2.1 were used for this approach. (Note that this is the method used in the second part of the example in Chapter 2 — Section 2.2.2.2.) Again, actual test data was used in both cases so that a direct comparison between the two methods could be made.

When the random uncertainty limit of C_d was calculated by the propagation method (method 1), it was assumed that only P_1, P_2, and T_1 significantly contributed to the random uncertainty:

$$P_{C_d} = P_{C_d}\left(P_1, P_2, T_1\right) \tag{6.2}$$

The random uncertainty associated with the mass flow rate determined by the standard was assumed to be negligible relative to the random uncertainties of the test venturi measurements of pressure and temperature. The standard deviations, S, of the three measurements (P_1, P_2, and T_1) were calculated from the 10 data scans for each test run (N=10). The random uncertainty limit for each of these mean quantities was then calculated using $P=2S/(10)^{1/2}$. The values of P_1, P_2, and T_1 were set to the average of the 10 scans for each point. These average values and the random uncertainty limits were propagated through the DRE to obtain an estimate of the random uncertainty limit for C_d. A central difference numerical scheme was used to approximate the partial derivatives. No correlation of the random errors in the measured variables was considered.

A summary of the results using the propagation method (method 1) is given in Table 6.2. These results indicate that the goal of achieving an uncertainty in C_d of 0.25% cannot be attained. Notice particularly the extremely high random uncertainties calculated for the low Mach number points.

Table 6.2 — C_d Random Uncertainty Limit Results

			P_{C_d} %	
M	Re x 10⁻⁶	C_d	Method 1	Method 2
0.197	0.975	0.9899	2.00	0.12
0.192	0.966	0.9917	4.37	0.11
0.201	1.084	0.9866	2.15	0.043
0.199	2.932	0.9894	2.46	0.082
0.199	5.953	0.9933	1.17	0.062
0.502	0.977	0.9885	0.37	0.050
0.495	2.971	0.9914	0.75	0.053
0.493	5.834	0.9927	0.14	0.061
0.698	1.450	0.9907	0.09	0.047
0.695	2.951	0.9914	0.23	0.028
0.683	5.894	0.9938	0.07	0.024

The random uncertainty limit of the discharge coefficient was then determined directly from the C_d results (method 2). C_d was calculated for each data scan, and the sample standard deviation, S, was then computed from the 10 values of C_d (N=10). The random uncertainty limit was again calculated using $P=2S/(10)^{1/2}$. These results are also given in Table 6.2. The results indicate that the goal of an uncertainty in C_d of 0.25% may be attained.

The random uncertainty limits calculated by method 1 were much higher than those calculated by method 2. The differences were especially large at the low Mach number points. Figure 6.2 helps explain this difference. The figure is a plot of the venturi inlet and throat pressures normalized to the critical flow meter inlet total pressure for a particular run. It shows that both P_1 and P_2 vary significantly with time; therefore, the random uncertainty limits for these two variables are large. Propagating these large random uncertainty limits using method 1 results in large random uncertainty limits for the discharge coefficient. However, notice that the ΔP remains fairly constant with time. It is the ΔP, not the absolute pressures P_1 and P_2, that is important for determining the discharge coefficient. The propagation method treated the random errors in the two venturi pressures as independent, but Fig. 6.2 shows that the variations of the two pressures are not independent. This same trend was seen for all of the test conditions. The fact that the throat pressure varied with the inlet pressure was a function of the test facility control. The distance between the critical flow meter and the venturi was small; therefore, pressure variations in the critical flow meter and in the venturi were in phase. The variations in the pressure measurements were not truly random; they were correlated. This correlation was not accounted for in the propagation analysis. Method 2 automatically accounted for the correlation effect since C_d was calculated for each scan, and this data was used to calculate the random uncertainty limit. To account for the correlation effect in the propagation analysis, the covariance, P_{ik}, must be estimated. To use this method for pretest predictions, one must recognize that the random errors of the two pressures will be correlated and include the correlation terms in the analysis.

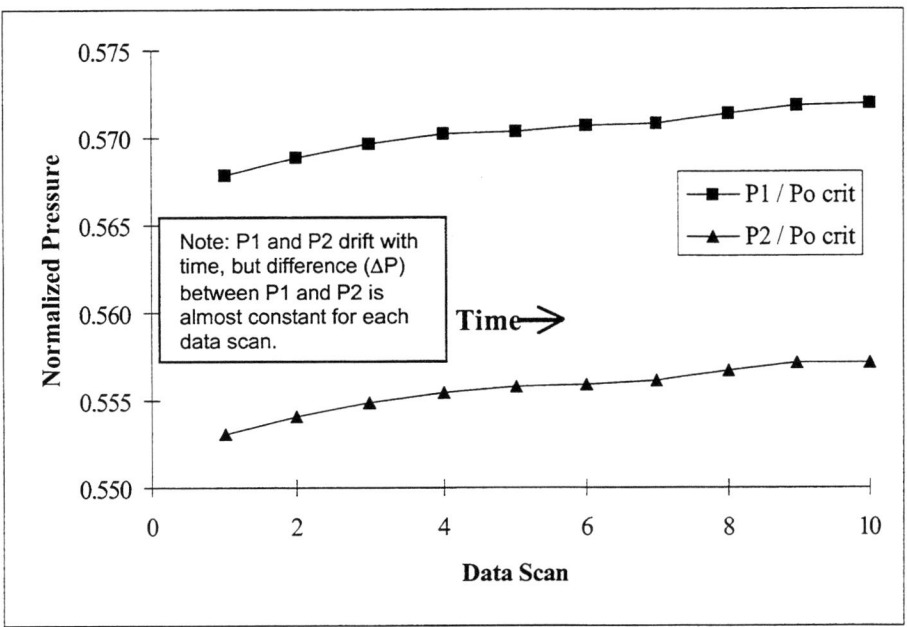

Figure 6.2 — Normalized Venturi Inlet and Throat Pressures

When test data is available, the random uncertainty limit should be calculated using method 2 to obtain the correct results. Calculating the random uncertainty limit by both methods is also recommended. Having the random uncertainty limits calculated by both methods would allow one to make a comparison between the two methods. If the random uncertainty limits from the two methods are significantly different, then the possibility of having correlated random uncertainty effects should be evaluated. This type of information can be important for planning similar future tests since method 1 must be used for pretest analyses when no data is available yet. Also note that method 2 cannot be used when the measurements are not made simultaneously. This leads to the conclusion that simultaneous measurements are preferred.

To gain more insight into the random uncertainty of the "worst case" point of M=0.2 and Re=1x10^6, three runs were made at this test condition. Two of the runs (10 scans each) were made at different times on the same day, and the third run (9 scans) was made almost two weeks later. The data for each run was calculated by method 2 as discussed previously. Combining the three runs gave N=29 for the overall data set. This data is summarized in Table 6.3. The data shows that the run-to-run repeatability is good. Some variation in C_d is expected due to set point variations. The random uncertainty limit of the overall average discharge coefficient is 0.10% when all three runs are considered (P=2S/(29)$^{1/2}$). This data set allows one to be more sure of the random uncertainty estimate for the "worst case" point as well as the other test points.

Table 6.3 — Venturi Calibration Multiple Test Results

Run / # of Scans	M	Re x 10^{-6}	C_d	P_{Cd} %
1 / 10	0.197	0.975	0.9899	0.12
2 / 10	0.192	0.966	0.9917	0.11
3 / 9	0.201	1.084	0.9866	0.043
Overall	0.197	1.008	0.9895	0.10

To summarize the random uncertainty analysis, the correlated random uncertainty terms can have a significant effect on the random uncertainty limit of a result. Random uncertainty estimates using propagation techniques without considering correlated random uncertainty terms were an order of magnitude larger than random uncertainty estimates calculated directly from a sample of results obtained at the same experimental set point. The potential effect of correlated random uncertainty terms makes it better to use multiple results rather than the propagation method to determine a random uncertainty limit for a result. The propagation method is useful in the planning phase, but one must recognize the effect of the correlated random uncertainty terms and include this in the analysis if it is applicable in a particular test situation. When the random uncertainty limit is automatically computed in the data reduction program, it should be computed by both methods so that a comparison can be made. If the random uncertainty limits are significantly different, then the possibility of having correlated random uncertainty effects should be considered. This information can then be useful in planning a similar experiment since the propagation method must be used in the planning phase. The correlated random uncertainty effect is probably much more prevalent than considered in the past since it occurs in any system in which multiple variables "drift" in unison in response to unsteadiness during the testing period.

To complete the discharge coefficient uncertainty analysis, the systematic uncertainty terms must be considered. Information provided by the calibration test group was used to determine the systematic uncertainty limit for each test point. A systematic uncertainty due to the curve-fit used to apply the calibration data was also estimated. These systematic uncertainty estimates were then combined with the random uncertainty estimates to obtain an overall uncertainty estimate at each test point (Table 6.4). The goal of $U_{Cd}=\pm0.25\%$ was achieved for all of the set points except M=0.2. The highest uncertainty was for the low Mach number and low Reynolds number point, as expected. The goal of $U_{Cd}=\pm0.25\%$ was not met for M=0.2, but the numbers were much lower than the pretest predictions.

Table 6.4 — Venturi Calibration Uncertainty Results

M	Re x 10^{-6}	P_{Cd} %	B_{Cd} %	B_{CF}	C_d	U_{Cd}	U_{Cd} %
0.197	1.008	0.10	0.450	0.0022	0.9895	0.0051	0.51
0.199	2.932	0.082	0.308	0.0022	0.9894	0.0038	0.39
0.199	5.953	0.062	0.391	0.0022	0.9933	0.0045	0.45
0.502	0.977	0.050	0.162	0.0015	0.9885	0.0022	0.23
0.495	2.971	0.053	0.134	0.0015	0.9914	0.0021	0.21
0.493	5.834	0.061	0.126	0.0015	0.9927	0.0020	0.21
0.698	1.450	0.047	0.138	0.0009	0.9907	0.0017	0.17
0.695	2.951	0.028	0.129	0.0009	0.9914	0.0016	0.16
0.683	5.894	0.024	0.128	0.0009	0.9938	0.0016	0.16

6.3 Discharge Coefficient and Mass Flow Rate Equations

The equations used to calculate the venturi Mach number, Reynolds number, mass flow rate, and discharge coefficient are included here. The equations are written for English units; therefore, pressure must be in psia, area in square inches, and temperature in °R. Constants used include $\gamma=1.4$ and R=53.35 ft-lb_f/lb_m°R. The constant 9.9×10^{-6} in Equation 7 is the thermal expansion coefficient.

(1) Venturi diameter ratio and pressure ratio

$$DR = \frac{d_2}{d_1} \text{ and } PR = \frac{P_2}{P_1} = \frac{P_1 - \Delta P}{P_1}$$

(2) Throat Mach number

$$M_2 = \sqrt{5\left[\frac{(PR)^{-2/7} - 1}{1-(DR)^4 (PR)^{10/7}}\right]}$$

(3) Inlet Mach number

$$M_1 = DR^2 PR^{6/7} M_2$$

(4) Inlet static temperature

$$T_1 = \frac{T_{01}}{1+0.2M_1^2}$$

(5) Inlet density

$$\rho_1 = \frac{144 P_1}{R T_1}$$

(6) Throat adiabatic wall temperature

$$T_{2aw} = T_{01}\left[\frac{1+0.1784 M_2^2}{1+0.2 M_2^2}\right]$$

(7) Throat diameter corrected due to difference between operating temperature and room temperature

$$d_{2_T} = d_2\left[1 + 9.9 \times 10^{-6}(T_{2aw} - 529.67)\right]$$

(8) Throat area based on throat diameter in (7)

$$A_{2_T} = \pi\left(\frac{d_{2_T}}{2}\right)^2$$

(9) Throat Reynolds number per foot per psia

$$\frac{\text{Re}}{dP_{02}} = \frac{(181{,}176{,}192) M_2\left[T_{01} + 198.6\left(1+0.2M_2^2\right)\right]}{(T_{01})^2\left(1+0.2M_2^2\right)^{2.5}}$$

(10) Throat Reynolds number

$$\text{Re} = \text{Re}_2 = \left(\frac{\text{Re}}{dP_{02}}\right) P_2 \left(1+0.2 M_2^2\right)^{3.5} \left(\frac{d_{2_T}}{12}\right)$$

(11) Ideal mass flow rate (lb$_m$/sec)

$$\dot{W}_{ideal} = \dot{W}_2 = 1.250605 (A_{2_T})\left[\frac{P_1 \rho_1 PR^{10/7}\left(1 - PR^{2/7}\right)}{1 - DR^4 PR^{10/7}}\right]^{1/2}$$

(12) Discharge coefficient (if calibration)

$$C_d = \frac{\dot{W}_{standard}}{\dot{W}_{ideal}} = \frac{\dot{W}_{actual}}{\dot{W}_{ideal}}$$

(13) Actual mass flow rate (if test)

$$\dot{W}_{actual} = C_d \dot{W}_{ideal}$$

References

6.1 Hudson, S.T., Bordelon, W.J. Jr., and Coleman, H.W., "Effect of Correlated Precision Errors on Uncertainty of a Subsonic Venturi Calibration," *AIAA Journal*, Vol. 34, No. 9, 1996.

6.2 Coleman, H.W. and Steele, W.G., "Experimentation and Uncertainty Analysis for Engineers," 2nd Edition, John Wiley & Sons, 1999.

6.3 Smith, R.E. and Matz, R.J., "A Theoretical Method of Determining Discharge Coefficients for Venturis Operating at Critical Flow Conditions," Transactions of the ASME, *Journal of Basic Engineering*, December 1962.

6.4 Lahti, D.J., "Theory and Experiments on Subcritical Compressible Gas Flow Metering," Ph.D. Dissertation, University of Cincinnati, Cincinnati, OH, August 1990.

6.5 Bean, H.S., (Ed.), "Fluid Meters, Their Theory and Application," ASME Report, 6th Edition, 1971.

6.6 Abernethy, R.B., Benedict, R.P., and Dowdell, R.B., "ASME Measurement Uncertainty," *J. Fluids Engineering*, Vol. 107, 1985.

6.7 American National Standards Institute/American Society of Mechanical Engineers, *Measurement Uncertainty*, PTC 19.1-1985 Part 1, ASME, 1986.

6.8 American National Standards Institute/American Society of Mechanical Engineers, *Measurement Uncertainty for Fluid Flow in Closed Conduits*, MFC-2M-1983, ASME, 1984.

7 Regression Uncertainty

Previous chapters in this Guide have focused on providing examples based on uncertainty methodology defined in the AIAA Standard [7.1]. This chapter now moves to a topic not covered in the Standard. Major advances have been made in the area of regression uncertainty since the Standard was originally published in 1995. These advances were underway while the Standard was being written; therefore, the material could not be included. Therefore, more detailed background information on the topic of regression uncertainty will be provided than has been typical of the previous chapters. Information in this chapter was compiled from references 7.2, 7.3, and 7.4. Further details may be found in those references. Reference 7.4 addresses the difference between the methodology presented here and the method previously used.

When experimental information is represented by a regression, or curve-fit (the terms can be used interchangeably for the purposes here), the regression model will have an associated uncertainty due to the uncertainty in the original experimental program. Determination of the uncertainty associated with a regression follows the same general approach as presented throughout this guide—the systematic and random uncertainties for the DRE are obtained by using the uncertainty propagation equations and combined using a root sum square.

Consider the general case where a test is conducted on a hardware component. The data is plotted and a linear regression is used to determine the best fit of a curve through the data. (Note that the term *linear* regression means the regression coefficients $a_0, a_1, ..., a_n$ are not functions of the X variable and does not mean the relationship between X and Y is linear.) Since the data is obtained experimentally, both X and Y will have experimental uncertainties, and these uncertainties will be made up of systematic (bias) and random (precision) uncertainties. A linear regression is then performed on the (X,Y) data, and the general form of the regression is

$$Y(X) = a_0 + a_1 X + a_2 X^2 + ... + a_n X^n \qquad (7.1)$$

This polynomial equation is then used to represent the performance of the hardware. The original data and its uncertainty are usually forgotten in subsequent use, and the uncertainties are fossilized into the curve-fit. This chapter presents a methodology to assess the uncertainty associated with regressions when the regression variables contain random, systematic, and correlated systematic uncertainties.

Obviously, an additional error is introduced if the wrong regression model is used; e.g., if a 3rd-order regression model is used and the true relationship is 1st-order. This is the classical problem of incorrectly fitting the data, overfitting or underfitting. <u>The error introduced by choice of an inappropriate regression model is not addressed here</u>. It is assumed in the techniques presented that the correct model is being used and the only uncertainties in the regression model are those due to the uncertainties in the original experimental information.

7.1 Categories of Regression Uncertainty

This methodology is developed to cover the types of situations commonly encountered in engineering for which regressions are used. Three primary categories can be defined based upon how the regression information is being used. The first category is when the regression coefficients are the primary interest. The second category is when the regression model is used to provide a value for the Y variable. And the third category is when the (X_i, Y_i) data points are not measured quantities, but are functions of other variables. This third category can include both of the other two categories, with the uncertainty associated with either the regression coefficients or the regression value of Y being of interest.

The key to the proper estimation of the uncertainty associated with a regression is a careful, comprehensive accounting of systematic and correlated systematic uncertainties. Correlated systematic uncertainties will be present and must be properly accounted for when the (X_i, Y_i) and the (X_{i+1}, Y_{i+1}) data pairs have systematic uncertainties from the same source. The examples presented below demonstrate some ways in which correlated systematic uncertainties can be present and must be properly accounted for.

7.1.1 Uncertainty in Coefficients

The general expression for a straight line regression is

$$Y(X) = mX + c \tag{7.2}$$

where m is the slope of the line and c is the y-intercept. In some experiments these coefficients are the desired information. An example is the stress-strain relationship for a linearly elastic material

$$\sigma = E\varepsilon \tag{7.3}$$

where the stress, σ, is linearly proportional to the strain, ε, by Young's modulus, E. Young's modulus for a material is determined by measuring the elongation of the material for an applied load, calculating the normal stress and strain, and determining the slope of the line in the linearly elastic region. The stress and strain will have experimental uncertainties since they are determined experimentally; therefore, the experimental value of Young's modulus will have an associated uncertainty.

7.1.2 Uncertainty in Y from Regression Model

Often, the uncertainty associated with a value determined using the regression model, Eq. (7.1) or (7.2), is desired. One would obtain a regression model for a given set of (X_i, Y_i) data, then use that regression model to obtain a Y value at a measured or specified X. The nature of the data and the use of the data determine how the uncertainty estimate is determined, such as

- Some or all (X_i, Y_i) data pairs from different experiments
- All (X_i, Y_i) data pairs from same experiment
- New X from same apparatus
- New X from different apparatus
- New X with no uncertainty

It is instructive to discuss an example of each of these situations.

Suppose heat transfer coefficient data sets from various facilities were combined as a single data set, with each facility contributing data over a slightly different range, and a regression model was generated. The random and systematic uncertainties for each test apparatus are different. If the systematic uncertainties for the (X_i, Y_i) data and the (X_{i+1}, Y_{i+1}) data are obtained from the same apparatus, and thus share the same error sources, their systematic uncertainties will be correlated. However, if they are from different apparati and do not share any error sources, they will not be correlated. The uncertainty associated with the regression model must properly account for the correlation of the systematic uncertainties.

The calibration of a thermocouple is an example where all of the data could come from the same experiment. A calibration curve for the thermocouple would be generated by measuring an applied temperature and an output voltage, both of which contain uncertainties. The calibration curve *could* have the form

$$T = mE + c \tag{7.4}$$

where m and c are the regression coefficients determined from the (E_i, T_i) calibration data. (Note that thermocouple calibrations are not always linear, particularly over wide temperature ranges.) When the thermocouple is then used in an experiment, a new voltage, E_{new}, is obtained. The new temperature is then found using the calibration curve

$$T_{new} = mE_{new} + c \tag{7.5}$$

The uncertainty in T_{new} includes the uncertainty in the calibration curve as well as the uncertainty in the voltage measurement, E_{new}. If the same voltmeter is used in the experiment as was used in the calibration, the systematic uncertainty from the new voltage measurement will be correlated with the systematic uncertainty of each E_i used in finding the regression and appropriate correlation terms are needed. If a different voltmeter is used to measure the new voltage, the systematic uncertainty of E_{new} will not be correlated with the systematic uncertainties of the E_i.

When a regression is used to represent a set of data and that regression is then used in an analysis, often the new X value is postulated and can be considered to have no uncertainty. An example would be pumping power, P, versus pump speed, N, for a centrifugal pump. The regression might have the form

$$P = a(N)^b \tag{7.6}$$

If an analyst uses this expression to obtain a value of power at a postulated value of N, then the uncertainty in N could be considered to be zero.

7.1.3 (X_i, Y_i) Variables are Functions

Another common situation is when the (X_i, Y_i) variables used in the regression are not the measured data but are each functions of several measured variables. The Young's modulus determination discussed previously is a typical example. Neither the stress nor strain is measured directly. The stress is calculated from measurement of the applied force with a load cell and measurement of the cross-sectional area. The DRE for stress is

$$\sigma = \frac{CV_{lc}}{V_i A} \tag{7.7}$$

where C is the calibration constant, V_{lc} is the load cell voltage, V_i is the excitation voltage, and A is the cross-sectional area. The strain is determined using a strain gage, and the DRE is

$$\varepsilon = \frac{2V_{br}}{GV_i} \tag{7.8}$$

where V_{br} is the bridge voltage, G is the gage factor, and again V_i is the excitation voltage. The regression coefficient representing Young's modulus is thus a function of the variables C, V_{lc}, V_i, V_{br}, A, and G. In instances where error sources are shared between different variables, as would exist if all of the voltages are measured with the same voltmeter, additional terms in the uncertainty propagation expression are necessary to properly account for the correlated systematic uncertainties.

7.2 Linear Regression Uncertainty

Linear regression analysis is based upon minimizing the sum of the squares of the Y-deviations between the line and the data points, commonly known as the method of least squares. Linear regression analysis can be divided into three broad categories: straight line regressions, polynomial regressions, and multivariate regressions. Multivariate regressions will not be discussed, but the technique should provide correct results if extended to them. Straight line regressions, also called 1st order regressions or simple linear regressions, are a commonly used form. It is often recommended that, if the data is not inherently linear, a transformation be used to try to obtain a linear relationship [7.5, 7.6]. Exponential functions and power law functions, for example, can be transformed to linear functions by appropriate logarithmic transformations. Reciprocal transformations are also very useful in linearizing a nonlinear function. If a suitable transformation cannot be found, a polynomial regression is then often used.

7.2.1 General Approach

As discussed earlier, the general approach to determining the uncertainty for a regression is to apply the uncertainty propagation equations to the regression function, $Y(X_i, Y_i)$, so that the uncertainty in each variable is propagated. The general expression for an nth order polynomial regression model is

$$Y(X_{new}) = a_0 + a_1 X_{new} + a_2 X^2_{new} \ldots + a_n X^n_{new} \tag{7.9}$$

where the regression coefficients, a_i, are determined with a least squares fit. A readily available numerical least squares regression method utilizes orthogonal polynomials and a singular decomposition solution routine and is described in Press[7.5]. A user should be able to use any standard numerical regression routine, spreadsheet or mathematical software function with this methodology. For higher order regressions ($n>1$) the complexity of the polynomial regression determination usually makes analytical determination of the partial derivatives prohibitive.

Considering Eq. (7.9) to be data reduction equations of the form

$$Y = Y(X_1, X_2, \ldots, X_N, Y_1, Y_2, \ldots, Y_N, X_{new}) \tag{7.10}$$

and applying the uncertainty analysis equations, the most general form of the expression for the uncertainty in the regression is

$$U_Y = \sqrt{B_Y^2 + P_Y^2} \tag{7.11}$$

where the systematic uncertainty is determined from

$$\begin{aligned} B_Y^2 &= \sum_{i=1}^{N} \left(\frac{\partial Y}{\partial Y_i}\right)^2 B_{Y_i}^2 + 2\sum_{i=1}^{N-1}\sum_{k=i+1}^{N} \left(\frac{\partial Y}{\partial Y_i}\right)\left(\frac{\partial Y}{\partial Y_k}\right) B_{Y_i Y_k} \\ &+ \sum_{i=1}^{N} \left(\frac{\partial Y}{\partial X_i}\right)^2 B_{X_i}^2 + 2\sum_{i=1}^{N-1}\sum_{k=i+1}^{N} \left(\frac{\partial Y}{\partial X_i}\right)\left(\frac{\partial Y}{\partial X_k}\right) B_{X_i X_k} \\ &+ 2\sum_{i=1}^{N}\sum_{k=1}^{N} \left(\frac{\partial Y}{\partial X_i}\right)\left(\frac{\partial Y}{\partial Y_k}\right) B_{X_i Y_k} \\ &+ \left(\frac{\partial Y}{\partial X_{new}}\right)^2 B_{X_{new}}^2 + 2\sum_{i=1}^{N} \left(\frac{\partial Y}{\partial X_{new}}\right)\left(\frac{\partial Y}{\partial X_i}\right) B_{X_{new} X_i} \\ &+ 2\sum_{i=1}^{N} \left(\frac{\partial Y}{\partial X_{new}}\right)\left(\frac{\partial Y}{\partial Y_i}\right) B_{X_{new} Y_i} \end{aligned} \tag{7.12}$$

The first five terms on the right hand side (RHS) of Eq. (7.12) account for uncertainties from the (X_i, Y_i) data pairs. The sixth and seventh terms account for the systematic uncertainty for the new X. The eighth term is included if the new X variable is measured with the same apparatus as that used to measure the original X variables to account for the correlated systematic uncertainty. The last term is included if error sources are common between the new X and the original Y_i variables.

If an analysis is being conducted using a Y value determined using Eq. (7.9) and it is postulated that the X value being used to evaluate the regression model has no uncertainty, then terms 6 through 8 on the RHS of Eq. (7.12) are omitted.

The random uncertainty, P_Y, is determined by propagating the random uncertainty associated with each variable using

$$P_Y^2 = \sum_{i=1}^{N} \left(\frac{\partial Y}{\partial Y_i}\right)^2 P_{Y_i}^2 + \sum_{i=1}^{N} \left(\frac{\partial Y}{\partial X_i}\right)^2 P_{X_i}^2 + \left(\frac{\partial Y}{\partial X_{new}}\right)^2 P_{X_{new}}^2 \tag{7.13}$$

where P_{Y_i} is the random uncertainty for the Y_i variable, and P_{X_i} is the random uncertainty for the X_i variable. Equation 7.13 provides the most general method to determine the random uncertainty, P_Y, associated with the regression; however, it can often be determined using standard regression analysis methods.

Using classical regression analysis techniques to determine the random uncertainty for use in Equation 7.11 is a valid option and may be more convenient. One simply replaces the P_Y^2 term with the equivalent 95% confidence random uncertainty interval determined from the statistical methods. These statistical methods are presented in section 7.3.1 for first-order regressions. Natrella, Montgomery and Peck, Press, and similar texts are good references for these methods. It should be noted that these methods are only valid, in a rigorous mathematical/statistical sense, when the X variables have no error or negligible error. It is often assumed that the random error in the X variables is transposed and captured in random variation of the Y variable. The methodology development work has verified this approach only for 1st order linear regressions. While similar verification work has not been conducted with higher order regressions, the extension appears to be valid.

7.2.2 Reporting Regression Uncertainties

After the uncertainty associated with a regression has been calculated, it should be documented clearly and concisely so it can be easily used. Figure 7.1 shows a set of (X,Y) data, the 1st order regression model for that data set, and the associated uncertainty interval determined using Eqs. (7.11-7.13). It would usually be very useful to have not only the regression model but also an equation giving $U_Y(X)$. While Eq. (7.11) gives such information, it requires having the entire (X_i, Y_i) data set as input each time a calculation of U_Y is made for a new X value. Since this is cumbersome and inconvenient, it is proposed that a set of (X, U) points be calculated using Eq. (7.11) and a regression be performed using these points to produce an expression for $U_{Y\text{-}regress}(X)$, which is used as explained below to calculate $U_Y(X)$ at a given X value.

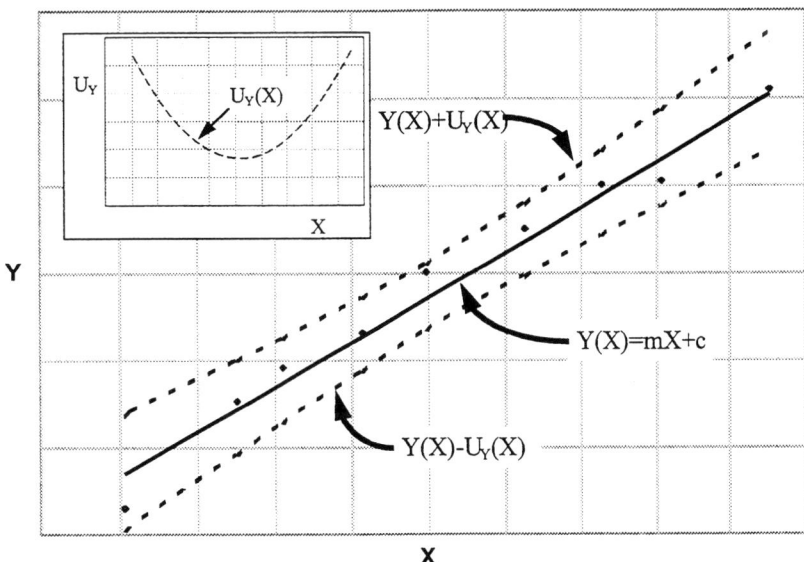

Figure 7.1 Expressing Regression Uncertainty Intervals

The procedure that should be used is as follows:

1. If the X_{new} to be used in Eq. (7.1) has no uncertainty, the last three terms of Eq.(7.12) and the last term of Eq. (7.13) are set equal to zero. The desired range of X variable values are used to produce the (X, U_Y) data set for the $U_Y(X)$ curvefit.

2. If the X_{new} to be used in Eq. (7.1) is from the same apparatus as produced the (X_i, Y_i) data set, then all of the terms in Eqs. (7.12 and 7.13) must be considered. However, the $P_{X_{new}}^2$ term will often be larger during the test than during the more well-controlled calibration. This can be taken into account by assigning the same systematic uncertainty components that were in the X_i data to the X_{new}'s used

in Eq. (7.11) to produce the (X, U_Y) points to curvefit, but with the $P^2_{X_{new}}$ term in Eq. (7.13) set equal to zero. The resulting (X, U_Y) points are then curvefit to yield $U_{Y\text{-}regress}(X)$. When an X_{new} is used in Eq. (7.1), the uncertainty U_Y in the Y determined from the regression is then calculated as

$$(U_Y)^2 = U^2_{Y-regress} + \left(\frac{\partial Y}{\partial X_{new}}\right)^2 P^2_{X_{new}} \qquad (7.14)$$

3. If the X_{new} to be used in Eq. (1) is from a different apparatus than that which produced the (X_i, Y_i) data set, then the last two terms in Eq. (7.12) will be zero. This situation would be encountered if a thermocouple calibration curve was determined but then another voltmeter was used in testing than that which produced the original (T_i, E_i) data set. Eq. (7.12) - with the last two correlated bias terms set equal to zero - is used over the desired X range to calculate a set of (X, U) points that are then curvefit to produce a $U_{Y\text{-}regress}(X)$ expression. When an X_{new} is used in Eq. (7.1), the uncertainty U_Y in the predicted Y is then calculated using

$$(U_Y)^2 = U^2_{Y-regress} + \left(\frac{\partial Y}{\partial X_{new}}\right)^2 \left[B^2_{X_{new}} + P^2_{X_{new}}\right] \qquad (7.15)$$

The contribution of $U_{Y\text{-}regress}$ to U_Y in Eqs. (7.14) and (7.15) is purely a systematic uncertainty, since all random uncertainties are fossilized once the regression is performed. It should be noted that the lowest-order curvefit that provides an acceptable fit for the regression uncertainty should be used to represent that uncertainty.

7.2.3 Differential Pressure Transducer Calibration Example

A differential pressure transducer was calibrated by applying a water column to each port of the transducer and recording the output voltage as the height of the water column on one side was increased. This differential pressure transducer was used to measure the differential pressure across the ports of a venturi for the determination of the water flowrate through a venturi. The differential pressure transducer calibration curve generated is shown in Figure 7.2.

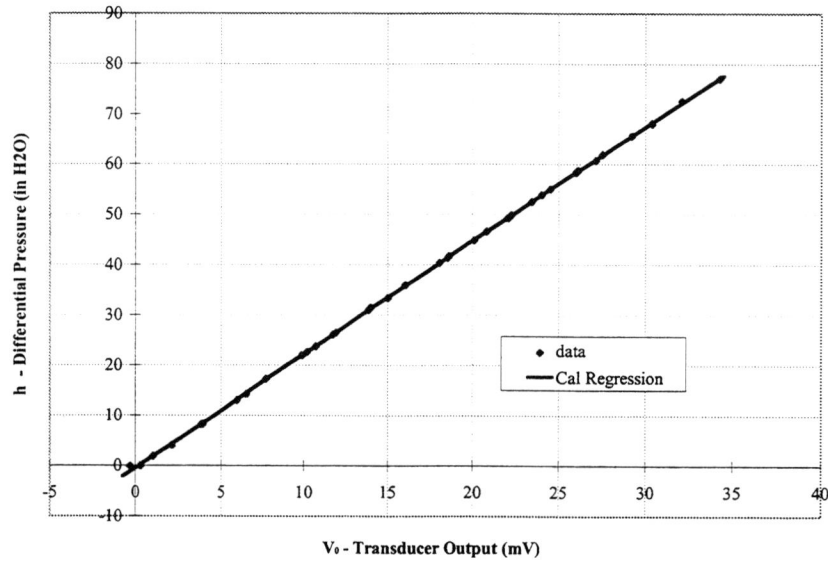

Figure 7.2 — Differential Pressure Transducer Calibration Curve

Using a 1st-order regression, the calibration curve for the differential pressure transducer was determined as

$$h = 2.2624V_o - 0.42334 \tag{7.16}$$

The uncertainty associated with the differential pressure transducer calibration curve was assessed by applying Equations 7.11 through 7.13 (where, in this case, $Y=h$ and $X=V_0$). No error sources were shared between the measured differential pressure and the measured transducer output voltage; therefore, the $B_{X_{new}Y}$ terms were not included. Since a random uncertainty estimate for X_{new} based upon the transducer calibration data might not be representative of the random uncertainty in X_{new} encountered when the transducer is used to determine the flowrate through the venturi, the $P_{X_{new}}$ term was omitted.

Hence, the random uncertainty for the transducer output voltage obtained during a test will have to be included in the final DRE uncertainty determination. This is a common situation since the calibration process is usually more well-controlled and provides more stable data than when the transducer is used in the actual experiment. The uncertainty for the predicted value of h from the differential pressure transducer calibration curve was written as

$$\begin{aligned}U_h^2 &= \sum_{i=1}^{N}\left(\frac{\partial h}{\partial h_i}\right)^2 P_{h_i}^2 + \sum_{i=1}^{N}\left(\frac{\partial h}{\partial V_{0i}}\right)^2 P_{V_{0i}}^2 \\&+ \sum_{i=1}^{N}\left(\frac{\partial h}{\partial h_i}\right)^2 B_{h_i}^2 + 2\sum_{i=1}^{N-1}\sum_{k=i+1}^{N}\left(\frac{\partial h}{\partial h_i}\right)\left(\frac{\partial h}{\partial h_k}\right)B_{h_i h_k} \\&+ \sum_{i=1}^{N}\left(\frac{\partial h}{\partial V_{0i}}\right)^2 B_{V_0(i)}^2 + 2\sum_{i=1}^{N-1}\sum_{k=i+1}^{N}\left(\frac{\partial h}{\partial V_{0i}}\right)\left(\frac{\partial h}{\partial V_{0k}}\right)B_{V_{0i}V_{0k}} \\&+ \left(\frac{\partial h}{\partial V_{0new}}\right)^2 B_{V_{0new}}^2 + 2\sum_{i=1}^{N}\left(\frac{\partial h}{\partial V_{0new}}\right)\left(\frac{\partial h}{\partial V_{0i}}\right)B_{V_{0new}V_{0i}}\end{aligned} \tag{7.17}$$

where the differential pressure, measured in inches of water column, h, was used as the Y variable and the transducer output voltage, V_0, in millivolts, was used as the X variable in the regression. A systematic uncertainty of 1/16" H_2O and a random uncertainty of 1/8" H_2O were used as the uncertainty estimates for the applied differential pressure. A systematic uncertainty estimate of 0.5 mV was used for the voltmeter based upon manufacturer specifications. The partial derivatives can be determined numerically using a finite difference technique or can be determined analytically. For example, the partial derivative of the regression with respect to the independent variables, X_i,

$$\frac{\partial Y}{\partial X_i} = \frac{\partial m}{\partial X_i}X + \frac{\partial c}{\partial X_i} \tag{7.18}$$

becomes

$$\frac{\partial h}{\partial V_0} = \frac{\partial m}{\partial V_0}V_0 + \frac{\partial c}{\partial V_0} \tag{7.19}$$

and the partial with respect to the dependent variable, Y_i,

$$\frac{\partial Y}{\partial Y_i} = \frac{\partial m}{\partial Y_i}X + \frac{\partial c}{\partial Y_i} \tag{7.20}$$

becomes

$$\frac{\partial h}{\partial h_i} = \frac{\partial m}{\partial h_i}V_0 + \frac{\partial c}{\partial h_i} \tag{7.21}$$

The partial derivatives are calculated at each V_0 for which the uncertainty of the calibration curve is desired. The analytically determined partial derivative, $\frac{\partial m}{\partial Y_i}$, (Eq. 7.43 in section 7.3) becomes

$$\frac{\partial m}{\partial h_i} = \frac{NV_{0i} - \sum_{i=1}^{N} V_{0i}}{N\sum_{i=1}^{N}(V_{0i}^2) - \left(\sum_{i=1}^{N} V_{0i}\right)^2} \tag{7.22}$$

and similarly, $\dfrac{\partial m}{\partial X_i}$ (Eq. 7.45) becomes

$$\frac{\partial m}{\partial V_{0_i}} = \frac{NY_i - \sum_{i=1}^{N} h_i}{N\sum_{i=1}^{N}(V_{0i}^2) - \left(\sum_{i=1}^{N} V_{0i}\right)^2} - \frac{\left(N\sum_{i=1}^{N} V_{0i} h_i - \sum_{i=1}^{N} V_{0i} \sum_{i=1}^{N} h_i\right)\left(2NV_{0i} - 2\sum_{i=1}^{N} V_{0i}\right)}{\left(N\sum_{i=1}^{N}(V_{0i}^2) - \left(\sum_{i=1}^{N} V_{0i}\right)^2\right)^2} \tag{7.23}$$

The uncertainty associated with the differential pressure transducer calibration curve was obtained from Equation 7.17 and is shown as a function of the output voltage in Figure 7.3.

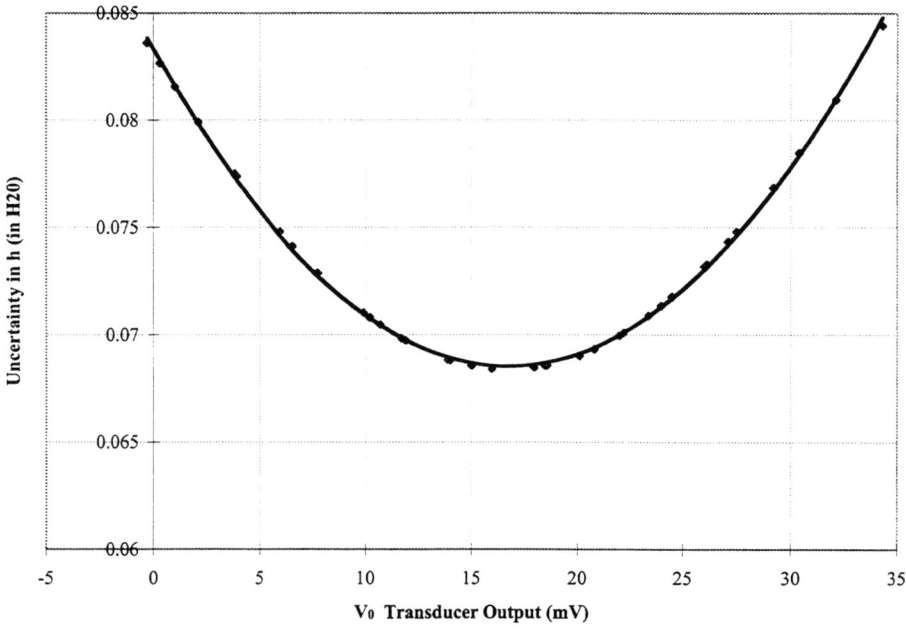

Figure 7.3 — Differential Pressure Transducer Calibration Curve Uncertainty

The 2nd order polynomial

$$U_h = 5.28 \times 10^{-5} V_o^2 - 0.00177 V_o + 0.0833 \tag{7.24}$$

was fit to the uncertainty results and was used to express the differential pressure uncertainty as a function of the transducer output voltage.

7.2.4 X and Y as Functional Relations

In many, if not most instances, the test results will be expressed in terms of functional relations, often dimensionless. In these cases, the measured variables will not be the X's and Y's used in the regression. Examples of common functional relations are Reynolds number, flow coefficient, turbine efficiency, specific fuel consumption, etc. For example, in the experimental determination of the pressure distribution on the surface of a cylinder in cross-flow, the results are usually presented as pressure coefficient versus angular position θ. The pressure coefficient is

$$C_P = \frac{P - P_\infty}{q_\infty} \qquad (7.25)$$

where P is the local pressure, P_∞ is the free-stream pressure, and q_∞ is the dynamic pressure. The Y_i variables used in the regression model are the determined pressure coefficient values and are functions of the three measured variables

$$Y_i = C_{p,i} = f(P_i, P_{\infty,i}, q_{\infty,i}) \qquad (7.26)$$

and the X_i variables are the angular positions θ_i. If the pressures are measured with separate transducers that were calibrated against the same standard, the systematic uncertainties from the calibration will be correlated and must be properly taken into account. Equation (7.12) *{and upcoming equations (7.30), (7.38), and (7.41)}* applies in such cases, with Y (in this case C_p) taken as the dependent function for the regression and the X_i's and Y_i's replaced by the full set of measured variables (in this case the P_i's, $P_{\infty,i}$'s, $q_{\infty,i}$'s, and θ_i's, assuming no uncertainty in the new value of θ.

In general, for an experiment with n sets of J measured variables, the regression model can be viewed as being a function of the J variables, such as

$$Y(X_{new}, X_i, Y_i) = f(VAR1_i, VAR2_i, \cdots, VARJ_i) \qquad (7.27)$$

where the function X_i is made up of some of the variables and the function Y_i is made up of some of the variables, such as

$$X_i = f(VAR1_i, VAR2_i, \cdots, VARK_i, \cdots) \qquad (7.28)$$

and

$$Y_i = f(\cdots, VARK_i, (VARK+1)_i, \cdots, VARJ_i) \qquad (7.29)$$

and X_{new} is the new value of the independent variable in the regression. The general expression for the uncertainty in the value determined from the regression model is again the RSS combination of the random and systematic uncertainties, Eq 7.11. The systematic uncertainty is determined from

$$
\begin{aligned}
B_Y^2 =\ & \sum_{i=1}^{n} \sum_{j=1}^{J} \left(\frac{\partial Y}{\partial VARj_i}\right)^2 B_{VARj_i}^2 \\
& + 2 \sum_{i=1}^{n-1} \sum_{k=i+1}^{n} \sum_{j=1}^{J} \left(\frac{\partial Y}{\partial VARj_i}\right)\left(\frac{\partial Y}{\partial VARj_k}\right) B_{VARj_i VARj_k} \\
& + 2 \sum_{i=1}^{n} \sum_{k=1}^{n} \sum_{j=1}^{J-1} \sum_{l=j+1}^{J} \left(\frac{\partial Y}{\partial VARj_i}\right)\left(\frac{\partial Y}{\partial VARl_k}\right) B_{VARj_i VARl_k} \\
& + \sum_{j=1}^{J} \left(\frac{\partial Y}{\partial VARj_{new}}\right)^2 B_{VARj_{new}}^2 + 2 \sum_{i=1}^{n} \sum_{j=1}^{J} \left(\frac{\partial Y}{\partial VARj_{new}}\right)\left(\frac{\partial Y}{\partial VARj_i}\right) B_{VARj_{new}} B_{VARj_i} \\
& + 2 \sum_{i=1}^{n} \sum_{j=1}^{J-1} \sum_{l=j+1}^{J} \left(\frac{\partial Y}{\partial VARj_{new}}\right)\left(\frac{\partial Y}{\partial VARl_i}\right) B_{VARj_{new}} B_{VARl_i}
\end{aligned}
\qquad (7.30)
$$

where the second RHS term accounts for correlated systematic uncertainty sources within each variable and the third RHS term accounts for systematic uncertainty sources common between variables. Similar expressions for the uncertainty in the slope and intercept are readily obtained. Since the X_i and Y_i variables used to determine the regression model are now functional relations, the determination of the partial derivatives becomes more complex and should most likely be performed numerically. The random uncertainty is determined from

$$P_Y^2 = \sum_{i=1}^{n} \sum_{j=1}^{J} \left(\frac{\partial Y}{\partial VARj_i} \right)^2 P_{VARj_i}^2 \qquad (7.31)$$

The partial derivatives can be approximated using a finite difference numerical technique, often referred to as a "jitter-routine." Round-off errors can be significant; therefore, double precision arithmetic may be helpful. Accuracy can also be enhanced by using a second-order accurate central-differencing scheme, such as:

$$\frac{\partial r}{\partial X_1} = \frac{r(X_1 + h, X_2, \ldots, X_n) - r(X_1 - h, X_2, \ldots, X_n)}{2h} \qquad (7.32)$$

Step sizes, h, on the order of 0.01% have proved to provide stable results for this type of use—smaller step sizes have a tendency to increase numerical accuracy problems, and larger step sizes can have errors when the functions have a high degree of local curvature.

7.2.5 1st Order Regressions

7.2.5.1 Uncertainty in Coefficients

The development of the equations to calculate m and c can be found in numerous statistics and regression books. [7.5, 7.6, 7.8] For N (X_i, Y_i) data pairs, the slope, m, is determined from

$$m = \frac{N \sum_{i=1}^{N} X_i Y_i - \sum_{i=1}^{N} X_i \sum_{i=1}^{N} Y_i}{N \sum_{i=1}^{N} (X_i^2) - \left(\sum_{i=1}^{N} X_i \right)^2} \qquad (7.33)$$

and the intercept is determined from

$$c = \frac{\sum_{i=1}^{N} (X_i^2) \sum_{i=1}^{N} Y_i - \sum_{i=1}^{N} X_i \sum_{i=1}^{N} (X_i Y_i)}{N \sum_{i=1}^{N} (X_i^2) - \left(\sum_{i=1}^{N} X_i \right)^2} \qquad (7.34)$$

Considering Eqs. (7.33) and (7.34) to be data reduction equations of the form

$$m = m(X_1, X_2, \ldots, X_N, Y_1, Y_2, \ldots, Y_N) \qquad (7.35)$$

and

$$c = c(X_1, X_2, \ldots, X_N, Y_1, Y_2, \ldots, Y_N) \qquad (7.36)$$

and applying the uncertainty analysis equations, the most general form of the expression for the uncertainty in the slope is

$$U_m = \sqrt{B_m^2 + P_m^2} \qquad (7.37)$$

where the systematic uncertainty for the slope is

$$B_m^2 = \sum_{i=1}^{N}\left(\frac{\partial m}{\partial Y_i}\right)^2 B_{Y_i}^2 + 2\sum_{i=1}^{N-1}\sum_{k=i+1}^{N}\left(\frac{\partial m}{\partial Y_i}\right)\left(\frac{\partial m}{\partial Y_k}\right)B_{Y_iY_k}$$
$$+ \sum_{i=1}^{N}\left(\frac{\partial m}{\partial X_i}\right)^2 B_{X_i}^2 + 2\sum_{i=1}^{N-1}\sum_{k=i+1}^{N}\left(\frac{\partial m}{\partial X_i}\right)\left(\frac{\partial m}{\partial X_k}\right)B_{X_iX_k} \qquad (7.38)$$
$$+ 2\sum_{i=1}^{N}\sum_{k=1}^{N}\left(\frac{\partial m}{\partial X_i}\right)\left(\frac{\partial m}{\partial Y_k}\right)B_{X_iY_k}$$

and the random uncertainty for the slope can be obtained by propagating the individual random uncertainties using

$$P_m^2 = \sum_{i=1}^{N}\left(\frac{\partial m}{\partial Y_i}\right)^2 P_{Y_i}^2 + \sum_{i=1}^{N}\left(\frac{\partial m}{\partial X_i}\right)^2 P_{X_i}^2 \qquad (7.39)$$

where

P_{Y_i} is the random uncertainty for the Y_i variable, P_{X_i} is the random uncertainty for the X_i variable,

B_{Y_i} is the systematic uncertainty for the Y_i variable,

B_{X_i} is the systematic uncertainty for the X_i variable,

$B_{Y_iY_k}$ is the covariance estimator for the correlated systematic uncertainties in the Y_i and Y_k variables, $B_{X_iX_k}$ is the covariance estimator for correlated systematic uncertainties in the X_i and X_k variables, and $B_{X_iY_i}$ is the covariance estimator for the correlated systematic uncertainties between X_i and Y_i.

A similar expression for the uncertainty in the intercept is

$$U_c^2 = B_c^2 + P_c^2 \qquad (7.40)$$

where the systematic uncertainty is

$$B_c^2 = \sum_{i=1}^{N}\left(\frac{\partial c}{\partial Y_i}\right)^2 B_{Y_i}^2 + 2\sum_{i=1}^{N-1}\sum_{k=i+1}^{N}\left(\frac{\partial c}{\partial Y_i}\right)\left(\frac{\partial c}{\partial Y_k}\right)B_{Y_iY_k}$$
$$+ \sum_{i=1}^{N}\left(\frac{\partial c}{\partial X_i}\right)^2 B_{X_i}^2 + 2\sum_{i=1}^{N-1}\sum_{k=i+1}^{N}\left(\frac{\partial c}{\partial X_i}\right)\left(\frac{\partial c}{\partial X_k}\right)B_{X_iX_k} \qquad (7.41)$$
$$+ 2\sum_{i=1}^{N}\sum_{k=1}^{N}\left(\frac{\partial c}{\partial X_i}\right)\left(\frac{\partial c}{\partial Y_k}\right)B_{X_iY_k}$$

and the random uncertainty is

$$P_c^2 = \sum_{i=1}^{N}\left(\frac{\partial c}{\partial Y_i}\right)^2 P_{Y_i}^2 + \sum_{i=1}^{N}\left(\frac{\partial c}{\partial X_i}\right)^2 P_{X_i}^2 \qquad (7.42)$$

These equations show the most general form of the expressions for the uncertainty in the slope and intercept, allowing for correlation of bias errors among the different X's, among the different Y's and also among the X's and Y's. If none of the systematic error sources are common between the X variables and the Y variables, the last term of Eqs. (7.38) and (7.41), the X-Y covariance estimator, is zero.

The partial derivatives are

$$\frac{\partial m}{\partial Y_i} = \frac{NX_i - \sum_{i=1}^{N} X_i}{N\sum_{i=1}^{N}(X_i^2) - \left(\sum_{i=1}^{N} X_i\right)^2} \quad (7.43)$$

$$\frac{\partial c}{\partial Y_i} = \frac{\sum_{i=1}^{N}(X_i^2) - X_i\sum_{i=1}^{N} X_i}{N\sum_{i=1}^{N}(X_i^2) - \left(\sum_{i=1}^{N} X_i\right)^2} \quad (7.44)$$

$$\frac{\partial m}{\partial X_i} = \frac{NY_i - \sum_{i=1}^{N} Y_i}{N\sum_{i=1}^{N}(X_i^2) - \left(\sum_{i=1}^{N} X_i\right)^2} - \frac{\left(N\sum_{i=1}^{N} X_i Y_i - \sum_{i=1}^{N} X_i \sum_{i=1}^{N} Y_i\right)\left(2NX_i - 2\sum_{i=1}^{N} X_i\right)}{\left(N\sum_{i=1}^{N}(X_i^2) - \left(\sum_{i=1}^{N} X_i\right)^2\right)^2} \quad (7.45)$$

and

$$\frac{\partial m}{\partial X_i} = \frac{2X_i\sum_{i=1}^{N} Y_i - \sum_{i=1}^{N} X_i Y_i - Y_i\sum_{i=1}^{N} X_i}{N\sum_{i=1}^{N}(X_i^2) - \left(\sum_{i=1}^{N} X_i\right)^2} - \frac{\left(\sum_{i=1}^{N}(X_i^2)\sum_{i=1}^{N} Y_i - \sum_{i=1}^{N} X_i \sum_{i=1}^{N} X_i Y_i\right)\left(2NX_i - 2\sum_{i=1}^{N} X_i\right)}{\left(N\sum_{i=1}^{N}(X_i^2) - \left(\sum_{i=1}^{N} X_i\right)^2\right)^2} \quad (7.46)$$

7.2.5.2 Classical Regression Random Uncertainty

As discussed previously, in some cases the random uncertainty can be determined using classical statistical regression techniques for the 95% confidence interval and combined with the systematic uncertainty using a RSS. The equations presented in this section are only valid for 1st order linear regressions.

The statistic that defines the standard deviation for a straight-line curvefit is the standard error of regression [7.6] defined as

$$S_Y = \left[\frac{\sum_{i=1}^{N}(Y_i - (mX_i + c))^2}{N-2}\right]^{\frac{1}{2}} \quad (7.47)$$

The regression equation given in Eq. (7.2) represents the relationship between the "mean" of Y and X. For a given value of X, the random uncertainty (95% confidence interval) associated with the "mean Y" obtained from the curve is

$$P_Y = 2\left(S_Y^2\left[\frac{1}{N} + \frac{(X - \overline{X})^2}{S_{XX}}\right]\right)^{\frac{1}{2}} \quad (7.48)$$

where

$$\overline{X} = \frac{1}{N}\sum_{i=1}^{N} X_i \quad (7.49)$$

and

$$S_{xx} = \sum_{i=1}^{N} X_i^2 - \frac{\left(\sum_{i=1}^{N} X_i\right)^2}{N} \qquad (7.50)$$

This random uncertainty interval around the curvefit value of Y for the given value of X should contain the parent population mean of Y with 95% confidence. In Equation 7.50, it is assumed that there are a large number of data pairs, N.

The statistical expressions for the random uncertainty of the slope and intercept are

$$P_m = 2S_m = 2\left[\frac{S_Y^2}{S_{XX}}\right]^{1/2} \qquad (7.51)$$

and

$$P_c = 2S_c = 2\left[S_Y^2\left(\frac{1}{N} + \frac{\overline{X}^2}{S_{XX}}\right)\right]^{\frac{1}{2}} \qquad (7.52)$$

7.2.5.3 Lift Slope and Lift Coefficient Example

As another simple example, consider the experimental determination of the lift slope for an airfoil. The lift slope is defined as the slope of the lift-coefficient, c_l, versus angle-of-attack, α, line

$$c_l = m\alpha + c \qquad (7.53)$$

As Anderson [7.9] states; "at low-to-moderate angles of attack, c_l varies *linearly* with α; the slope of this straight line is denoted by a_0 and is called the *lift slope*." (Note, while a_0 is the standard notation in aerodynamics literature for the lift slope, it will be defined as m in this work so that it will not be confused as a regression coefficient.) A thin, symmetric airfoil, such as a NACA 0012, will result in a theoretical lift slope of 2π

$$lift\ slope = m = \frac{dc_l}{d\alpha} = 2\pi \qquad (7.54)$$

when thin airfoil theory is applied [7.6]. The lift coefficient per unit width

$$c_l = \frac{l}{q_\infty L} = 2\pi\alpha \qquad (7.55)$$

is the lift force, l, divided by the free-stream dynamic pressure, q_∞, and the chord length, L. The dynamic pressure for incompressible flow,

$$q_\infty = \frac{1}{2}\rho U_\infty^2 = p_0 - p_\infty \qquad (7.56)$$

is the difference between the local total and static pressures and is measured with a Pitot-static probe.

The lift slope is a function of the lift force, the free-stream dynamic pressure, the chord length, L, and the angle of attack measurements, written functionally as

$$m = \frac{dc_l}{d\alpha} = f(l, q_\infty, L, \alpha) \qquad (7.57)$$

Applying the uncertainty propagation equations to Equation 7.57, the expression for the uncertainty associated with the lift slope is

$$U_m^2 = \sum_{i=1}^{N}\left(\frac{\partial m}{\partial l_i}\right)^2 B_l^2(i) + \sum_{i=1}^{N}\left(\frac{\partial m}{\partial q_i}\right)^2 B_{q_i}^2 + \sum_{i=1}^{N}\left(\frac{\partial m}{\partial \alpha_i}\right)^2 B_{\alpha_i}^2$$

$$+ 2\sum_{i=1}^{N-1}\sum_{k=i+1}^{N}\left(\frac{\partial m}{\partial l_i}\right)\left(\frac{\partial m}{\partial l_k}\right)B_{l_i l_k} + 2\sum_{i=1}^{N-1}\sum_{k=i+1}^{N}\left(\frac{\partial m}{\partial q_i}\right)\left(\frac{\partial m}{\partial q_k}\right)B_{q_i q_k} + 2\sum_{i=1}^{N-1}\sum_{k=i+1}^{N}\left(\frac{\partial m}{\partial \alpha_i}\right)\left(\frac{\partial m}{\partial \alpha_k}\right)B_{\alpha_i \alpha_k} \quad (7.58)$$

$$+ \sum_{i=1}^{N}\left(\frac{\partial m}{\partial l_i}\right)^2 P_{l_i}^2 + \sum_{i=1}^{N}\left(\frac{\partial m}{\partial q_i}\right)^2 P_{q_i}^2 + \sum_{i=1}^{N}\left(\frac{\partial m}{\partial \alpha_i}\right)^2 P_{\alpha_i}^2$$

where the middle line of Equation 7.58 accounts for the correlated systematic uncertainties within each variable, and the systematic and random uncertainty associated with the airfoil length is negligible. (Note: similar expressions can be obtained for the uncertainty associated with the intercept.) Using the analytical form of the partial derivatives, as shown in Eq. 7.32 and Eq. 7.34, differentiation can be accomplished using the chain rule; however, in general, it is not recommended due to complexities that arise in the correlated systematic uncertainty terms when the same variable is present in both the independent and dependent variable functions. Treating the lift slope as a function of the data variables - lift force, l, the dynamic pressure, q, and the angle of angle-of-attack, α - the partial derivatives can be determined numerically as discussed at the end of section 7.1. Each variable is "jittered" to determine the partial derivative of the lift slope, m, with respect to that variable. Three examples are shown in the following equations:

$$\frac{\partial m}{\partial l_{i=1}} = \frac{1}{2\Delta l}\begin{pmatrix} m(l_1 + \Delta l, l_2, \ldots, l_n, q_1, q_2, \ldots, q_n, \alpha_1, \alpha_2, \ldots, \alpha_n) \\ - m(l_1 - \Delta l, l_2, \ldots, l_n, q_1, q_2, \ldots, q_n, \alpha_1, \alpha_2, \ldots, \alpha_n) \end{pmatrix} \quad (7.59)$$

$$\frac{\partial m}{\partial q_{i=2}} = \frac{1}{2\Delta q}\begin{pmatrix} m(l_1, l_2, \ldots, l_n, q_1, q_2 + \Delta q, \ldots, q_n, \alpha_1, \alpha_2, \ldots \alpha_n) \\ - m(l_1, l_2, \ldots, l_n, q_1, q_2 - \Delta q, \ldots, q_n, \alpha_1, \alpha_2, \ldots, \alpha_n) \end{pmatrix} \quad (7.60)$$

$$\frac{\partial m}{\partial \alpha_{i=n}} = \frac{1}{2\Delta \alpha}\begin{pmatrix} m(l_1, l_2, \ldots, l_n, q_1, q_2, \ldots, q_n, \alpha_1, \alpha_2, \ldots \alpha_n + \Delta \alpha) \\ - m(l_1, l_2, \ldots, l_n, q_1, q_2, \ldots, q_n, \alpha_1, \alpha_2, \ldots, \alpha_n - \Delta \alpha) \end{pmatrix} \quad (7.61)$$

The lift slope uncertainty was determined by developing an equation similar to Equation 7.11 through 7.13 for Equation 7.53, since the Y-axis, dependent variable, is a function of other variables. The uncertainty associated with values predicted using the regression model, the lift coefficient equation, is

$$U_{c_l}^2 = \sum_{i=1}^{N}\left(\frac{\partial c_l}{\partial l_i}\right)^2 B_l^2(i) + \sum_{i=1}^{N}\left(\frac{\partial c_l}{\partial q_i}\right)^2 B_{q_i}^2 + \sum_{i=1}^{N}\left(\frac{\partial c_l}{\partial \alpha_i}\right)^2 B_{\alpha_i}^2$$

$$+ 2\sum_{i=1}^{N-1}\sum_{k=i+1}^{N}\left(\frac{\partial c_l}{\partial l_i}\right)\left(\frac{\partial c_l}{\partial l_k}\right)B_{l_i l_k} + 2\sum_{i=1}^{N-1}\sum_{k=i+1}^{N}\left(\frac{\partial c_l}{\partial q_i}\right)\left(\frac{\partial c_l}{\partial q_k}\right)B_{q_i q_k} + 2\sum_{i=1}^{N-1}\sum_{k=i+1}^{N}\left(\frac{\partial c_l}{\partial \alpha_i}\right)\left(\frac{\partial c_l}{\partial \alpha_k}\right)B_{\alpha_i \alpha_k} \quad (7.62)$$

$$+ \sum_{i=1}^{N}\left(\frac{\partial c_l}{\partial l_i}\right)^2 P_{l_i}^2 + \sum_{i=1}^{N}\left(\frac{\partial c_l}{\partial q_i}\right)^2 P_{q_i}^2 + \sum_{i=1}^{N}\left(\frac{\partial c_l}{\partial \alpha_i}\right)^2 P_{\alpha_i}^2$$

where the partial derivatives are again determined using the finite-difference technique similar to Eqs. 7.59-7.61.

The systematic uncertainty for the calibration of the pressure transducer was estimated to be 0.005 kPa based upon manufacturer specifications. The systematic uncertainty for the calibration of the lift force transducer was estimated to be 0.05 N based upon the transducer calibration. A correction of +3° was applied to all of the angle-of-attack measurements to correct for the flow angularity, and a systematic uncertainty of 0.5° was assigned to account for the uncertainty in the correction and the systematic uncertainty associated with the angle-of-attack measurement scale. A random uncertainty of 0.5° was estimated for the angle-of-attack measurements, based upon the readability of the angle-of-attack measurements. The free-stream velocity was set by selecting dynamic pressure setpoints at 0.1 kPa increments between 0.6 and 1.4 kPa. The random uncertainty estimate for the free-stream dynamic pressure measurements was estimated as 0.152 kPa, based upon a pooled standard deviation about the mean of the averaged data for each angle-of-attack setting. The random uncertainty for the force measurements was estimated based upon the standard deviation of the averaged data at each angle-of-attack setting. The random uncertainty estimate for each angle-of-attack setting is shown in Table 7.1. The uncertainty due to the wind tunnel data acquisition system was assumed negligible with respect to the magnitude of the other uncertainties. All random uncertainty estimates were made using the large sample approximation and the systematic and random uncertainty estimates are summarized in Table 7.1.

Table 7.1 — Uncertainties for Lift Slope Determination

Source	Systematic	Random
Angle-of-attack, a	0.5°	0.5°
Force, per unit span, l	0.05 N	Table 7.2
Dynamic Press, q_∞	0.005 kPa	0.152 kPa

Test data was obtained for 7-10 different dynamic pressure set-points at each angle-of-attack setting, and the corresponding lift force was recorded. These values were averaged to obtain the data set used in the regression, as shown in Table 7.2, and the random uncertainty was calculated using the standard deviation of the data at each set-point.

Table 7.2 — Lift Slope Determination Experimental Data and Lift Random Uncertainties

α (degrees)	q_∞ (kPa)	l (N)	P_l (N)
-10.0	0.949	-27.05	5.10
-6.0	0.951	-17.80	3.30
-3.5	1.048	-11.45	1.87
-0.5	1.051	-0.375	0.12
3.0	1.016	10.075	1.87
5.0	1.153	15.838	2.12
7.0	1.053	21.35	3.40
9.0	1.249	30.513	3.7

The lift slope and the associated uncertainty for the NACA 0012 airfoil tested is

$$m = 0.095 \pm 0.010 \tag{7.63}$$

and the intercept and its associated uncertainty is

$$c = -0.0067 \pm 0.07 \tag{7.64}$$

The regression model to predict the lift coefficient as a function of the angle-of-attack for the airfoil tested is

$$c_l(\alpha) = 0.095\alpha - 0.0067 \tag{7.65}$$

with the uncertainty associated with the lift coefficient model, U_{c_l}, given by the curvefit expression

$$U_{c_l} = 0.0008\alpha^2 - 0.0019\alpha + 0.0599 \tag{7.66}$$

Figure 7.4 is a plot of the lift coefficient versus angle-of-attack. It includes the test data, the regression through the data, the associated uncertainty interval, and the theoretical lift slope line.

The theoretical lift slope of 2π/rad equates to 0.1097/degree, which is not within the interval around the experimental slope as provided in Eq. 7.61. It is also seen in Figure 7.4 that the theoretical performance of a NACA 0012 airfoil is not entirely within the uncertainty interval about the regression model for the test data over the entire range of α. This deviation is a natural consequence of comparing experimental data to a theoretical prediction. The difference is explainable by noting that conceptual systematic uncertainties are present which are not included in the uncertainty calculations. For example, the theoretical prediction is based upon potential flow theory, which does not include consideration of viscous effects, wall effects, turbulence intensity, Reynolds number effects, etc., but the experimental results are affected by these considerations.

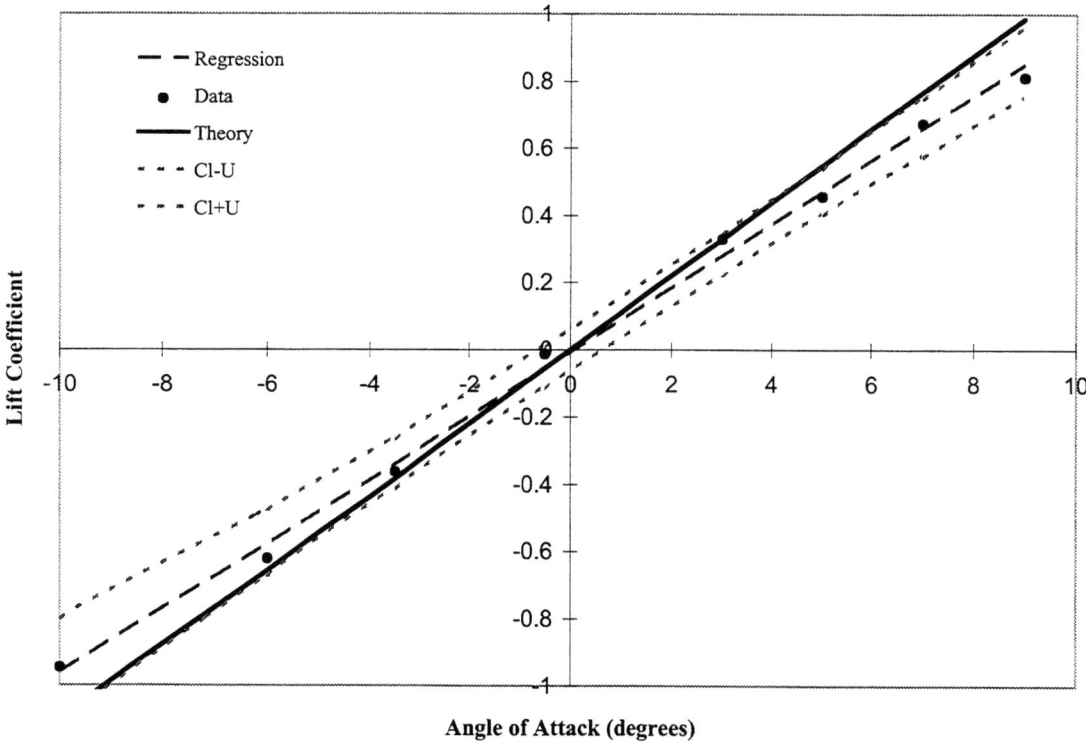

Figure 7.4 — Lift Coefficient vs Angle-of-Attack for a NACA 0012 Airfoil

References

7.1 "Assessment of Experimental Uncertainty with Application to Wind Tunnel Testing," AIAA Standard S-071A-1999.

7.2 Brown, Kendall K., 1996, "Assessment of the Experimental Uncertainty Associated with Regressions," Ph.D. Dissertation, University of Alabama in Huntsville, Huntsville, AL.

7.3 Brown, Kendall K., Coleman, Hugh W., Steele, W. Glenn, "A Methodology for Determining Experimental Uncertainties in Regressions," *Journal of Fluids Engineering,* Vol. 120, No. 3, 1998, pp. 445-456.

7.4 Coleman, H.W. and Steele, W.G., "*Experimentation and Uncertainty Analysis for Engineers,*" 2nd Edition, John Wiley & Sons, 1999.

7.5 Seber, G.A.F., 1977, *Linear Regression Analysis*, John Wiley & Sons, New York.

7.6 Montgomery, Douglas C. and Peck, Elizabeth A., 1992, *Introduction to Linear Regression Analysis*, 2nd Ed., John Wiley & Sons, New York.

7.7 Press, William H., Flannery, Brian P., Teukolsky, and Vetterling, William T., 1986, *Numerical Recipes: The Art of Scientific Computing*, Cambridge University Press, Cambridge.

7.8 Natrella, M.G., 1963, *Experimental Statistics*, NBS Handbook 91, National Bureau of Standards, United States Department of Commerce.

7.9 Anderson, John D., 1984, *Fundamentals of Aerodynamics*, McGraw-Hill, New York.

8 Automated Uncertainty Analysis for Production Experiments

8.1 Introduction

As we have seen, uncertainty analysis in the planning phase of an experiment provides a rational basis for the choices of experimental technique and instrumentation. Post test, an uncertainty analysis helps to quantify the quality actually achieved. Conceptually, this is obvious. The challenge is to perform the analysis for any and all experimental results of interest to the researcher or customer and to do it for any and all test conditions. For the typical wind-tunnel test, the task can be daunting.

In the first place, the experimental domain is often large and complex. Data may be acquired for several configurations of the test article, over a range of Mach numbers/dynamic pressures, and at a variety of model attitudes. The partial derivatives that describe the sensitivities of the results to the inputs, which are required for error propagation, can be complex functions to evaluate. In general, they are not constant over the experimental domain. Also, the uncertainties of the inputs themselves may vary with tunnel configuration, with model configuration, or even with measured value. These considerations are multiplied for each experimental result, of which there are many: freestream conditions, aerodynamic angles, force and moment coefficients, centers of pressure, surface pressure coefficients, control-surface load coefficients, etc. It is, therefore, difficult to foresee which test points will be critical from the standpoint of uncertainty.

This chapter presents a detailed application of uncertainty analysis to a production wind tunnel test. A generic fighter-type model was tested in a blow-down wind tunnel at transonic Mach numbers from 0.45 to 1.6 and supersonic Mach numbers from 1.45 to 2.48. The model angle-of-attack was varied from -4 to +8 degrees. The emphasis is on the automation of procedures and the propagation of errors to provide the maximum understanding of the data flow. The techniques and approach discussed can be applied to most wind-tunnel facilities.

There are three main tasks in performing an uncertainty analysis for a production experiment. The first is to construct an error-propagation tool that is suited to the particular facility and database. This is critical if one wishes to analyze arbitrary test points on demand. The second task is to estimate the systematic and random uncertainties of all input parameters. This is typically the most time-consuming step. It involves collecting relevant information from a variety of sources, including metrology reports, operator calibrations, check-load results, and model design drawings. The specific testing procedures will need to be considered in order to properly define any correlations among the error sources. Finally, the third task is to report the results in a way that is understandable and useful to both the customer and the experimentalist.

8.2 Error Propagation

Error propagation involves the evaluation of partial derivatives and bookkeeping. The derivatives can be evaluated analytically, with finite differences ("jittering"), or by automatic differentiation [8.1]. To facilitate the goal of uncertainty analysis on demand, a propagation application should access the production database of input values and measurements. Finally, the application should output results that are meaningful and in a convenient format.

To appreciate the scale of the task, consider that this example requires at least 52 inputs:

- facility parameters consisting of the supply temperature and pressure, the transonic plenum pressure, and the supersonic freestream Mach number

- balance forces and moments

- primary model reference quantities consisting of the reference area, the span, and the chord

- model and balance moment-reference-center locations in the body axis system

- model quantities consisting of normal and axial areas plus o .set distances for each of the base and cavity pressure corrections
- 2 freestream flow-angularity corrections in pitch and yaw
- 2 commanded angles comprising the pitch and roll of the model positioning system
- 16 offset and deflection angles defining four coordinate system rotations
- 3 model-to-balance offset angles
- 2 or more base and cavity pressure transducers

There are also 30 results of interest:

- freestream flow properties consisting of the Mach number, dynamic pressure, static temperature, static pressure, density, velocity, mass flux ,and Reynolds number
- aerodynamic angles consisting of the angles-of-attack and sideslip in the stability axes, the total angle-of-attack, the angle-of-sideslip in the body axes, and the aerodynamic roll angle
- force and moment coefficients in the body-axes system
- force and moment coefficients in the stability-axes system, corrected for base and cavity pressure
- 1 lift-to-drag ratio
- 2 base and cavity pressure coefficients
- 2 center-of-pressure locations for pitch and yaw

The analysis of a single test point will require the evaluation of 30_52 =1560 partial derivatives. The data reduction equations will not be reproduced here. Suffice it to say the production computer code consists of approximately 900 lines. In order to keep the problem manageable and reduce the opportunity for error, a propagation application employing finite differences for the differentiation was developed. The program is based on the flowchart given on page 190 of Reference [8.2]. It consists of a generic main program and two problem-specific modules. The main code performs the differentiation, manages the bookkeeping, and outputs the results. The first module defines the inputs and results of interest, reads the input uncertainties from a user-generated file, and interfaces with the production database of measured values and constants. The second module implements the data reduction equations. The code is validated by confirming that the calculated nominal values of the results agree with the production database. Experience has shown that fairly rapid implementations are possible. Also, the temptation to perform the analysis with respect to anything other than the independent variables has been removed.

The propagation application can generate up to three types of output. The first output type summarizes the systematic, random, and total uncertainties of all the results for a single test point. The second output type tabulates the uncertainties for multiple test points; it captures the variation of uncertainty with measured value. This format is most useful to customers and researchers for plotting purposes. The final output type reports the contribution of the uncertainty in each measured input to the overall uncertainty of each computed result. This report is most revealing to the experimentalist and test facility personnel. These three output types are referred to as summaries, tabulations, and reports, respectively.

8.3 Estimation of Input Uncertainties

In order to apply the methodology consistently and routinely to all measurements, the following general procedures are observed. The basic philosophy is to perform an end-to-end assessment, as opposed to accumulating the effects of many elemental error sources. This minimizes bookkeeping and, hopefully, maximizes the objectivity. It is assumed that any requirements for traceability to national standards have been met.

8.3.1 Systematic Uncertainty Estimation

When estimating systematic errors, the inputs generally fall into one of three classes. The first class of input is a constant, such as a reference area or an offset distance. In this case, all relevant information will have to be considered. For example, when considering the reference length of the test article, machining tolerances can serve as the systematic estimate. Other specific examples will be given in a later section. One should not be discouraged in these situations, but should make an estimate and continue with the analysis. The estimate can always be refined should such an input prove to be a dominant contributor to the overall uncertainty of a result.

The second class includes those quantities measured with a device whose uncertainty is provided by a metrology organization. One example is the tunnel supply pressure transducer, which is calibrated in-place by the local metrology laboratory. Metrology reports typically give a single accuracy specification. In the absence of further classification information, there is no choice but to take this value as representing a systematic uncertainty:

$$B = B_{metrology} \tag{8.1}$$

The confidence level associated with the value can usually be ascertained from the report. The term *standard uncertainty* indicates a one-sigma estimate, while the term *expanded uncertainty* indicates a two-sigma or 95% confidence level [8.3, 8.4].

The third class of input involves those instruments calibrated by the end users. Force balances and model pressure transducers are common examples. There are at least two contributors to systematic uncertainty. The first contributor is the working standard against which the instrument is calibrated, B_{ws}. Again, this value is obtained directly from the relevant metrology report. The second contributor to the systematic uncertainty is the calibration process, which is typically described by least-squares regressions. If the regression model is appropriate and the residuals are randomly distributed with a constant variance, an interval equal to twice the standard error of regression will, in general, encompass 95% of the calibraton data points. Therefore the calibration systematic uncertainty is taken to be $B_{cal} = 2S_Y$ [8.5, 8.2], where the standard error of regression, S_Y, is given by Equation 7.47 for a straight-line fit. For a general case, the standard error of regression is given by:

$$S_Y = \left[\sum_{i=1}^{N} \frac{R_i^2}{N-c} \right]^{1/2} \tag{8.2}$$

where N is the number of points in the regression, R_i are the regression residuals, and c is the number of coefficients determined by the regression. The total systematic uncertainty estimate is then

$$B = \left[B_{ws}^2 + B_{cal}^2 \right]^{1/2}$$
$$B = \left[B_{ws}^2 + (2S_Y)^2 \right]^{1/2} \tag{8.3}$$

Sometimes one must consider other systematic uncertainty contributors. The most common instances are when measurements fall outside of the calibration range, or when the calibration is performed in a separate apparatus. Only a thorough review of the particular circumstances can determine whether additional systematic terms should be added to Equation 3. For the present example, no additional contributions were deemed significant.

8.3.2 Random Uncertainty Estimation

In general, one must consider covariances among the inputs when estimating random uncertainties. The present approach chooses to propagate only those random uncertainties that are genuinely random. This avoids special analyses to define the correlation coefficients and simplifies the bookkeeping in the error propagation code. Random uncertainties for error propagation, then, are estimated from in-place readings using the entire measurement system as $P = 2S$. Here, S is the standard deviation, and the factor 2 indicates that the large sample assumption has been made.

The standard deviations were computed from tare data recorded before and after each tunnel blow under quiescent conditions. For N tunnel blows, there will be $2N$ realizations of S for each input channel. The 95th percentile of the $2N$ realizations is computed as a conservative estimate of S. This is an instrumentation-level evaluation that defines the system noise floor. It is sometimes referred to as a *zeroth replication level* evaluation [8.6, 8.7]. This random uncertainty is often negligible compared with the systematic uncertainties. Additional sources of variation that affect within-test and test-to-test repeatability, which are not included in the above estimates, must be assessed from appropriate replicate test points. Past experience can provide some guidance if such repeat runs are not included in the test matrix.

8.3.3 Specific Uncertainty Estimates

Uncertainties for the input measurements and parameters were made according to the above guidelines, and are presented in Table 8.1. The following notes and assumptions apply to this analysis:

- A systematic uncertainty of 0.005 in was assumed for the model span b and mean aerodynamic chord c, based on model design tolerances. This value was then propagated to the model reference area through the equation $S = b \cdot c$.

- A systematic uncertainty of 0.003 in was assumed for the balance-to-model transfer distances, based on model design tolerances.

- A systematic uncertainty of 0.01 deg was assumed for the balance-to-model offset angles.

- The uncertainty in the model cavity area was estimated by assuming an uncertainty of 0.003 in on the equivalent radius.

- Flow angularity corrections were determined for each blow and are assumed to have zero uncertainty.

- Deflections of the support system arc sector were assumed negligible, with zero uncertainty.

- All support system angles, including sting/balance deflections and model-to-balance offsets, were measured with or calibrated against a single servo inclinometer. This gives rise to 28 correlated systematic pairs, all of which are equal to the uncertainty of the device (0.02 deg provided by local metrology).

- The two cavity pressure transducers were calibrated against a common reference standard, and therefore have a correlated systematic uncertainty equal to the uncertainty in the reference standard.

Table 8.1 also lists uncertainties in the transonic freestream static-to-total pressure ratio and the supersonic set-point Mach number. Physically, these values are the variation in centerline tunnel conditions as measured in facility calibrations. They are treated as independent systematic uncertainties on the freestream conditions in addition to that arising from the instrumentation used in the facility calibration [8.8]. Attention is drawn to this quantity since, at some operating conditions; it can be a significant contributor to the overall uncertainty in the computed results.

8.4 Results

Some typical results will now be presented. To aid in interpretation, the nomenclature for measured inputs and calculated results are given in Tables 8.2 and 8.3. Twenty-four runs were analyzed, representing 16 test conditions in the transonic cart and eight test conditions in the supersonic test section. A run consisted of an angle-of-attack sweep from approximately -4 to +8 degrees, captured in 60-65 scans. Every fourth scan was analyzed for tabular output. Summary and report outputs, being more voluminous, were generated for every 15th scan. Execution time for a single run was less than five seconds on a PentiumIII-class desktop computer. The computations were automated using the `make` utility, allowing the entire analysis to be updated in a few minutes as uncertainty estimates evolved or new data became available.

Excerpts of results for a run at Mach 0.85 are presented in Fig. 8.1-8.3. Figure 8.1 shows an excerpt from the summary file, which is a snapshot of all results for a single scan of data.

Uncertainty budgets for drag coefficient are shown in Fig. 8.2 and 8.3. Contributions are listed as a percentage of the maximum contribution. A value of 100 therefore identifies the largest source of uncertainty. This is a slight variation of the UPC defined in Eq. 2.16, which is normalized by the total uncertainty. Figure 8.2 is for scan 16, which occurred at a low angle of attack of −1.1 degrees. One can see that the balance axial force (`AF`) dominates the uncertainty. The next largest contributor is the pitch deflection of the sting (`dadefl`), followed by the indicated sector position (`alphai`), and several correlated systematic uncertainty terms. But these other contributors are essentially negligible, since the percentages given are "under the square root". For example, the `alphai` systematic contribution is calculated as $\left[(6/100) \cdot (0.0002869)^2\right]^{1/2} = 0.00007$, or less than 1 drag count. Their cumulative root-sum-squared effect brings the total systematic to 3.3 counts, which is less than 0.5 counts over the base contribution from `AF` of 2.9 counts. Random terms, being estimated at such a low replication level, are indeed negligible compared with the systematic uncertainties.

The picture changes at higher pitch attitudes. Figure 8.3 shows the uncertainty budget for drag at an angle of attack of +5.3 degrees. One can see that the dominant source of uncertainty is now the pitch deflection of the sting (`dadefl`), which by itself contributes 4 counts to a total of 7.3. This is followed by the indicated sector position (`alphai`) and the balance axial force (`AF`). Note also that correlated systematic terms contribute much more significantly.

Examination of the uncertainty budgets helps the experimentalist prioritize improvements to the system. If minimum drag is of concern (low angle of attack), the priority should be decreasing the uncertainty of the balance measurements. But if cruise drag is important, efforts should first be directed towards reducing the uncertainty in the angle-of-attack system, beginning with sting deflection corrections and indicated sector position. Consideration should also be given to using separate working standards to measure/characterize the angles `alphai`, `alphas`, and `dadefl`. This alone would eliminate the three correlated systematic terms, reducing the total uncertainty in Fig. 8.3 from 7.3 counts to 6.2 counts.

The tabulation output files allow one to construct plots such as those in Fig. 8.4 and 8.5, which again are for drag coefficient. The error bars in Fig. 8.4 show the uncertainty in relation to the nominal value over the sweep. Figure 8.5 explicitly shows that the uncertainty is a function of measured value, varying from 3 to 10 drag counts. Other types of analyses are also possible. For example, say one wished to examine how the uncertainty in drag coefficient varies across the Mach number tested. To see this, the tabulation files were interpolated at a value of `CDS` =0.0600 to produce Fig. 8.6. The uncertainty is higher at low Mach due to the lower tunnel dynamic pressures at these operating conditions.

8.5 Summary

This chapter presents a detailed application of uncertainty analysis to a production wind tunnel test. The emphasis is on the automation of procedures and the propagation of errors to provide the maximum understanding of the data flow. General guidelines for random and systematic error estimation are given, along with some specifics for an actual wind tunnel test. Sample results are plotted, and error budgets are examined. The uncertainty drivers were found to change with measurement condition. The techniques and approach can be applied to most wind-tunnel facilities.

References

8.1. Meyn, L.A., "A New Method for Integrating Uncertainty Analysis into Data Reduction Software." AIAA Paper 98-0632, 36th AIAA Aerospace Sciences Meeting and Exhibit, Reno, NV, January 1998.

8.2. Coleman, H.W. and Steele, W.G., *Experimentation and Uncertainty Analysis for Engineers*," 2nd Edition, John Wiley & Sons, 1999.

8.3. ISO/TAG/WG3. *Guide to the Expression of Uncertainty in Measurement*. First edition 1995. International Organization for Standardization, 1993.

8.4. Taylor, B.N. and Kuyatt, C.E., "Guidelines for Evaluating and Expressing the Uncertainty of NIST Measurement Results." Technical Note 1297, 1994 Edition, National Institute of Standards and Technology, September 1994.

8.5. Kammeyer, M.E., "Uncertainty Analysis for Force Testing in Production Wind Tunnels," Proceedings of the First International Symposium on Strain-Gauge Balances, NASA Langley Research Center, pages 221-242, October 1996.

8.6. Moffat, R.J., "Contributions to the Theory of Single-sample Uncertainty Analysis," *Journal of Fluids Engineering*, 104:250-260, June 1982.

8.7. Kammeyer, M.E. and Rueger, M.L., "On the Classification of Errors: Systematic, Random, and Replication Level," AIAA Paper 2000-2203, 21st AIAA Aerodynamic Measurement Technology and Ground Testing Conference, Denver, CO, June 2000.

8.8. Kammeyer, M.E., "Wind Tunnel Facility Calibrations and Experimental Uncertainty," AIAA Paper 98-2715, 20th AIAA Advanced Measurement and Ground Testing Conference, Albuquerque, NM, June 1998.

Table 8.1 — Estimated Uncertainties for Instrumentation and Input Parameters

Quantity (units)	Transducer Type	F.S Range or Meas. Value	Uncertainty Systematic	Random
Facility Instrumentation and Parameters				
Supply pressure PO (psia)	Paroscientific	45	0.009	0.007
		200	0.040	0.002
		400	0.080	0.004
Supply temp TO (°R)	Cu vs. Cu45Ni Thermocouple	1200	--	0.05
Transonic section plenum pressure PP (psia)	Kulite average of 4	50	0.01	0.001
Freestream static-to-total pressure ratio P/PO	Transonic facility calibration	@M=0.45	1.5×10^{-4}	0
		@M=0.60	2.5×10^{-4}	0
		@M=0.70	3.9×10^{-4}	0
		@M=0.75	3.5×10^{-4}	0
		@M=0.80	3.9×10^{-4}	0
		@M=0.85	4.0×10^{-4}	0
		@M=0.90	4.8×10^{-4}	0
		@M=0.95	4.7×10^{-4}	0
		@M=1.00	4.2×10^{-4}	0
		@M=1.05	8.0×10^{-4}	0
		@M=1.10	1.2×10^{-3}	0
		@M=1.20	1.8×10^{-3}	0
		@M=1.30	1.7×10^{-3}	0
		@M=1.45	1.2×10^{-3}	0
		@M=1.50	1.1×10^{-3}	0
		@M=1.60	1.1×10^{-3}	0
Set point Mach number M	Supersonic facility calibration	1.45	0.015	0
		1.50	0.015	0
		1.60	0.014	0
		1.70	0.013	0
		1.83	0.012	0
		1.97	0.010	0
		2.15	0.012	0
		2.25	0.013	0
Flow angularity angu (deg)	Decicated runs	Various	0	0
				0
Sector angles Alphai (deg) Omegia(deg)	Beckman Potentiometer	-20 to +20 -90 to +90	0.037 0.094	0.002 0.011
Sector-to-sting angles alphas (deg) psis (deg) Omegas (deg)	Schaevitz LSO-C-30 Servo-inclinometer	Various	0.02 0.02 0	0 0 0
Sting-to-balance angles alphas (deg) psisb (deg) omegasb(deg)		0 0 0	0 0 0	0 0 0

Table 8.1 (continued) — Estimated Uncertainties for Instrumentation and Input Parameters

Quantity (units)	Transducer Type	F.S Range or Meas. Value	Uncertainty Systematic	Uncertainty Random
Support system deflection angles				
dapod (deg)		0	0	0
dppod (deg)		0	0	0
dopod (deg)		0	0	0
Sting deflection angles	Calibrated versus Balance load	Calculated		
dadef1 (deg)			0.051	0
dpdef1 (deg)			0.049	0
dodef1 (deg)			0.234	0
Model instrumentation and parameters				
Cavity pressures	Kulite			
Pcav1 (psia)		50	0.025	0.001
Pcav2 (psia)		50	0.081	0.003
Internal balance loads	1.5 in MK 21 J force balance	Cal range		
FN (lbf)		2390	2.13	0.22
PM (in-lbf)		3985	3.50	0.50
SF (lbf)		1046	1.88	0.06
YM (in-lbf)		1250	4.61	0.14
RM (in-lbf)		1200	1.81	0.09
AF (lbf)		250	0.36	0.02
Balance moment ref. Center location				
fsb (in)		21.12	0	0
blb (in)		0	0	0
wlb (in)		5.00	0	0
Reference area / lengths				
s (ft^2)		1.040	8.3*10e-4	0
b (in)		22.894	0.005	0
cbar (in)		6.981	0.005	0
Cavity area / offset dist.				
sc (ft^2)		0.0187	1.2*10e-4	0
xc (in)		N/A	0	0
zc (in)		0	0	0
Model moment ref. Center location				
fsm (in)		19.68	0.003	0
blm (in)		0	0.003	0
wlm (in)		5.220	0.003	0
Balance-to-model angles				
im (deg)		0.0558	0.01	0
ip (deg)		-0.0015	0.01	0
omegamb (deg)		0	0.01	0

Table 8.2 — Nomenclature for Measured Inputs

P0	Tunnel supply pressure
TT0	Tunnel supply temperature
PP	Transonic Plenum pressure
P VAR	Uncertainty in tunnel calibrated static-to-total pressure ratio
MACH	Supersonic Mach number
M VAR	Uncertainty in tunnel calibrated supersonic Mach number
NF	Balance normal force
PM	Balance pitching moment
SF	Balance side force
YM	Balance yawing moment
RM	Balance rolling moment
AF	Balance axial force
S	Model reference area
b	Model span
cbar	Model chord
FSB	Balance moment reference center, fuselage station
BLB	Balance moment reference center, butt line
WLB	Balance moment reference center, water line
FSM	Model moment reference center, fuselage station
BLM	Model moment reference center, butt line
WLM	Model moment reference center, water line
angu	Flow angularity in tunnel pitch plane
pangu	Flow angularity in tunnel yaw plane
alphai	Measured sector pitch angle
psii	Measured sector yaw angle
omegai	Measured sector roll angle
alphapd	Roll pod offset pitch angle
psipd	Roll pod offset yaw angle
omegapd	Roll pod offset roll angle
dapod	Roll pod pitch deflection
dppod	Roll pod yaw deflection
dopod	Roll pod roll deflection
alphas	Sting offset pitch angle
psis	Sting offset yaw angle
omegas	Sting offset roll angle
dadefl	Sting pitch deflection
dpdefl	Sting yaw deflection
dodefl	Sting roll deflection
alphasb	Sting-to-balance pitch angle
psisb	Sting-to-balance yaw angle
omegasb	Sting-to-balance roll angle
im	Model-to-balance pitch offset angle
ip	Model-to-balance yaw offset angle
PcavXX	Cavity pressures

Table 8.3 — Nomenclature for Calculated Results

MACH	Freestream Mach number
Q	Freestream dynamic pressure
RN	Freestream Reynolds number
P	Freestream static pressure
V	Freestream velocity
RHO	Freestream density
TS	Freestream static temperature
RHOV	Freestream mass flux RHO_V
ALPHA	Angle of attack
ALPHAP	Total angle of attack
BETA	Angle of sideslip in the stability-axes system
BETAP	Angle of sideslip in the boxy-axes system
PHIP	Aerodynamic roll angle
CNFU	Body-axes normal force coefficient, uncorrected
CPMU	Body-axes pitching moment coefficient, uncorrected
CSFU	Body-axes side force coefficient, uncorrected
CYMU	Body-axes yawing moment coefficient, uncorrected
CRMU	Body-axes rolling moment coefficient, uncorrected
CAFU	Body-axes axial force coefficient, uncorrected
CLS	Stability-axes lift force coefficient, corrected for base/cavity pressure
CPMS	Stability-axes pitching moment coefficient, corrected for base/cavity pressure
CSFS	Stability-axes side force coefficient, corrected for base/cavity pressure
CYMS	Stability-axes yawing moment coefficient, corrected for base/cavity pressure
CRMS	Stability-axes rolling moment coefficient, corrected for base/cavity pressure
CDS	Stability-axes drag force coefficient, corrected for base/cavity pressure
L/D	Lift-to-Drag ratio
Cpcav	Cavity pressure coefficient
XCP	Center of pressure location, pitch
YCP	Center of pressure location, yaw

```
                    PB&J UNCERTAINTY PROPAGATION
          Run  41, Transonic test cart, Mach= 0.85, Config W/B/T

Point    46
         Result      Nom. value      Syst. unc.      Rand. unc.      Urss          Units

         MACH    ,    8.4810E-01,    9.011E-04,     7.763E-05,     9.044E-04,
         Q       ,    1.2242E+03,    1.645E+00,     1.381E-01,     1.651E+00,     psf
         RN      ,    7.5104E+06,    7.325E+04,     9.599E+02,     7.326E+04,     1/ft
         P       ,    1.6885E+01,    1.478E-02,     1.295E-03,     1.484E-02,     psia
         V       ,    9.1658E+02,    3.539E+00,     8.398E-02,     3.540E+00,     ft/s
         RHO     ,    2.9128E-03,    2.208E-05,     3.075E-07,     2.208E-05,     slg/ft3
         TS      ,    4.8647E+02,    3.681E+00,     4.521E-02,     3.681E+00,     degR
         RHOV    ,    2.6698E+00,    1.020E-02,     1.584E-04,     1.020E-02,     slg/ft2s
         ALPHA   ,    5.2760E+00,    8.362E-02,     1.799E-03,     8.364E-02,     deg
         ALPHAP  ,    5.2763E+00,    8.264E-02,     1.772E-03,     8.266E-02,     deg
         BETA    ,    4.8057E-02,    6.276E-02,     7.369E-04,     6.277E-02,     deg
         BETAP   ,    4.8262E-02,    6.303E-02,     7.401E-04,     6.303E-02,     deg
         PHIP    ,    5.2261E-01,    6.859E-01,     8.013E-03,     6.859E-01,     deg
         CNFU    ,    4.5415E-01,    1.818E-03,     1.801E-04,     1.827E-03,
         CPMU    ,   -1.7649E-03,    5.592E-04,     6.685E-05,     5.631E-04,
         CSFU    ,   -1.6877E-03,    1.481E-03,     4.744E-05,     1.482E-03,
         CYMU    ,    9.9118E-05,    1.834E-04,     5.570E-06,     1.835E-04,
         CRMU    ,   -3.7410E-04,    8.720E-05,     3.200E-06,     8.726E-05,
         CAFU    ,    6.0495E-03,    2.979E-04,     1.604E-05,     2.983E-04,
         CLS     ,    4.5167E-01,    1.812E-03,     1.794E-04,     1.821E-03,
         CPMS    ,   -1.7649E-03,    5.592E-04,     6.685E-05,     5.631E-04,
         CSFS    ,   -1.6877E-03,    1.481E-03,     4.744E-05,     1.482E-03,
         CYMS    ,    1.3310E-04,    1.821E-04,     5.532E-06,     1.822E-04,
         CRMS    ,   -3.6340E-04,    8.975E-05,     3.265E-06,     8.980E-05,
         CDS     ,    4.7785E-02,    7.338E-04,     2.699E-05,     7.343E-04,
         L/D     ,    9.4521E+00,    1.428E-01,     4.297E-03,     1.429E-01,
         Cp_cav  ,   -3.2540E-02,    5.321E-03,     2.601E-04,     5.327E-03,
         Cp_base ,    0.0000E+00,    0.000E+00,     0.000E+00,     0.000E+00,
         XCP     ,    1.9653E+01,    9.998E-03,     1.020E-03,     1.005E-02,     in
         YCP     ,    2.1025E+01,    2.146E+00,     6.386E-02,     2.147E+00,     in
```

Figure 8.1 — Excerpt from Summary Output File at MACH = 0.85

```
                     PB&J UNCERTAINTY PROPAGATION
           Run  41, Transonic test cart, Mach= 0.85, Config W/B/T

                            Result Summary

Scan    16
                      CDS      =  1.672E-02

                      Urss =   3.295E-04         ;   1.97%
                      B    =   3.290E-04         ;   1.97%
                      P    =   1.684E-05         ;   0.10%


                          CONTRIBUTION COMPARISON
                    as % of max contributor of  2.869E-04

               Input              Systematic term        Random term

               P0                      0.2                  0.0
               PP                      0.2                  0.0
               P_VAR                   0.2                  0.0
               NF                      1.4                  0.0
               AF                    100.0                  0.3
               S                       0.2                  0.0
               alphai                  6.0                  0.0
               alphas                  1.7                  0.0
               dadefl                 11.4                  0.0
               dodefl                  0.0                  0.0
               omegasb                 0.0                  0.0
               ip                      0.0                  0.0
                 .                      .
                 .                      .
                 .                      .

               Correlated systematic terms:
               alphai  ,alphas         3.4
               alphai  ,dadefl         3.4
               alphai  ,dodefl         0.0
               alphas  ,dadefl         3.4
               alphas  ,dodefl         0.0
               dadefl  ,dodefl         0.0
                 .                      .
                 .                      .
                 .                      .
```

Figure 8.2 — Excerpt from Report Output File at MACH = 0.85 (Contribution comparison at ALPHA = -1.10. Only the most significant contributors are shown.)

```
                    PB&J UNCERTAINTY PROPAGATION
          Run   41, Transonic test cart, Mach= 0.85, Config W/B/T

                            Result Summary

Scan     46
                    CDS     =   4.778E-02

                    Urss =   7.343E-04          ;    1.54%
                     B   =   7.338E-04          ;    1.54%
                     P   =   2.699E-05          ;    0.06%


            CONTRIBUTION COMPARISON
            as % of max contributor of  4.081E-04

            Input              Systematic term          Random term

            P0                        0.8                  0.0
            PP                        0.8                  0.0
            P_VAR                     0.9                  0.0
            NF                       13.8                  0.1
            AF                       49.0                  0.2
            S                         0.9                  0.0
            alphai                   53.0                  0.1
            alphas                   14.9                  0.0
            dadefl                  100.0                  0.0
              .                        .                    .
              .                        .                    .
              .                        .                    .

            Correlated systematic terms:
            alphai  ,alphas        29.8
            alphai  ,dadefl        29.8
            alphas  ,dadefl        29.8
              .                      .
              .                      .
              .                      .
```

Figure 8.3 — Excerpt from Report Output File at MACH = 0.85 (Contribution comparison at ALPHA = 5.276. Only the most significant contributors are shown.)

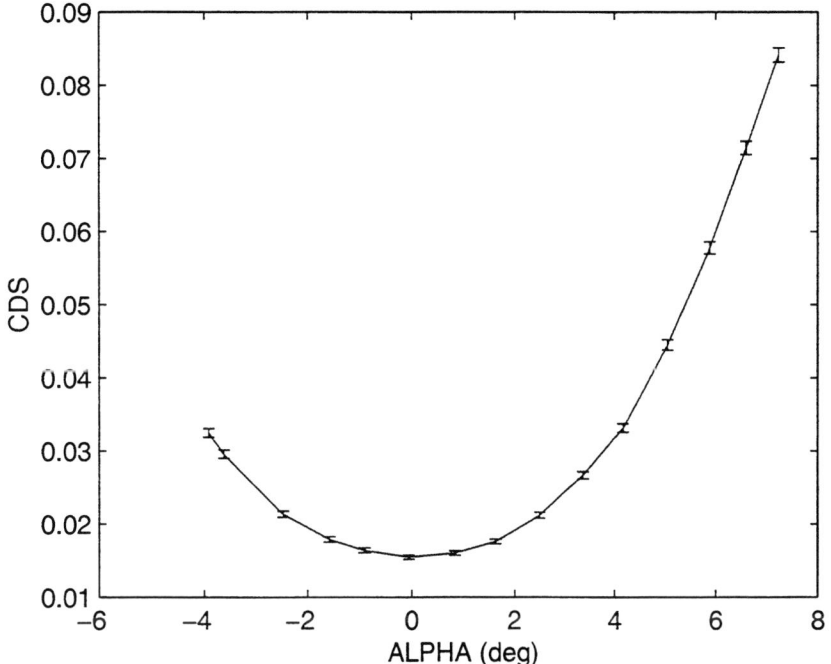

Figure 8.4 — Stability-axes Drag Coefficient at MACH= 0.85

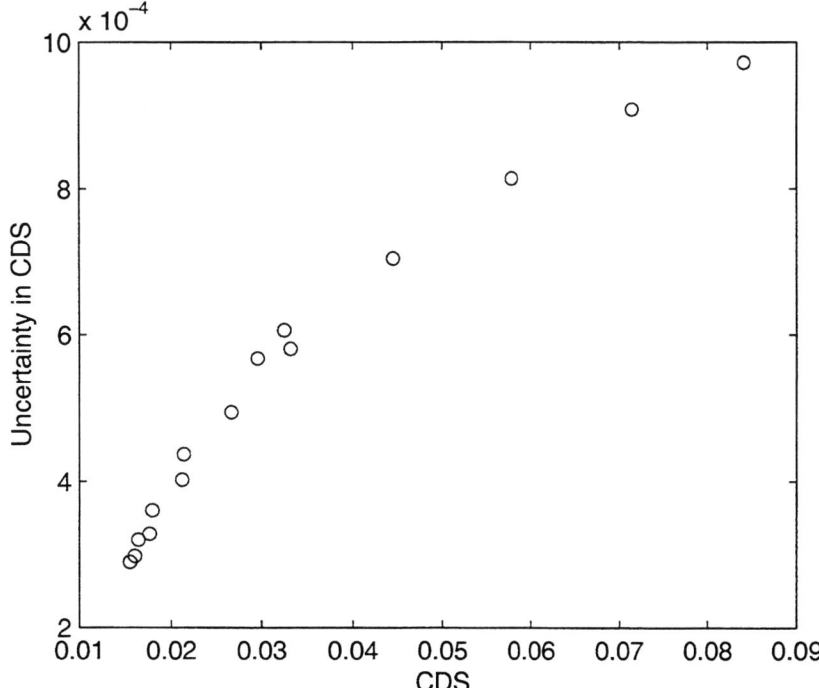

Figure 8.5 — Variation of Uncertainty in Stability-axes Drag Coefficient with Measured Value at MACH = 0.85

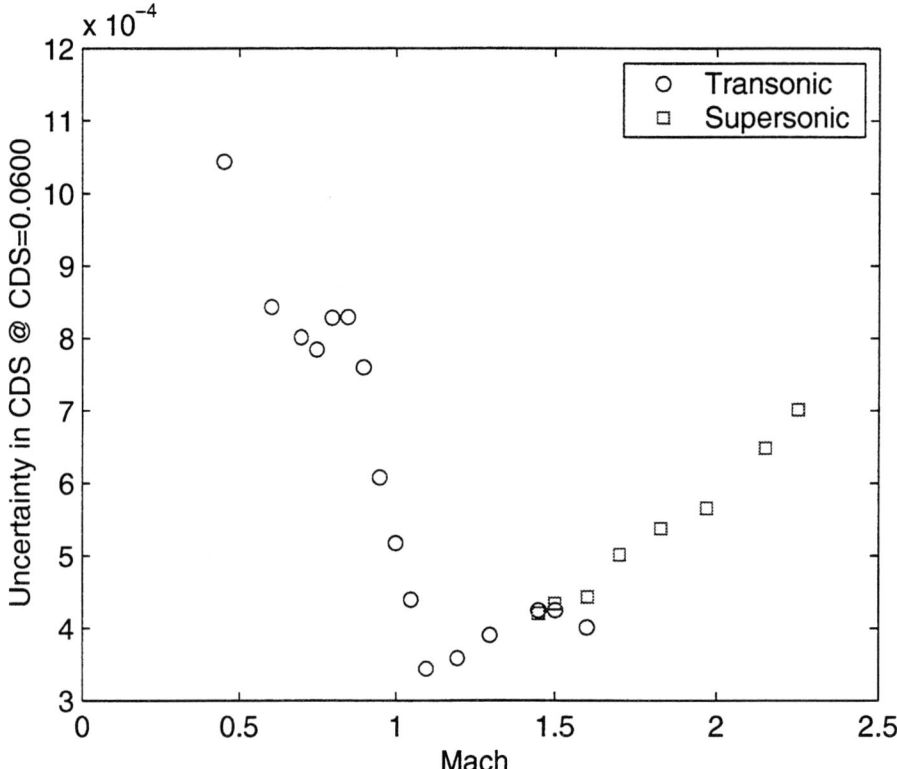

Figure 8.6 — Variation of Uncertainty in the Value CDS=0.0600 with Mach Number

9 Useful References

This chapter contains a list of references related to uncertainty analysis. The list was compiled by collecting input from GTTC members as well as others over the course of preparing this document. The subcommittee does not endorse any of these references as the "correct" method on any particular subject. The list is simply included to provide users with a head start on gathering information in their area of interest. Users are responsible for evaluating the merit of each publication.

[1] Zhang, Y., "Second-order Effects on Uncertainty Analysis Calculations," M.S. Thesis, Dept. of Mechanical Engineering, Mississippi State University, Starkville, MS, Dec. 2002.

[2] Hudson, S.T., Steele, W.G., Ryan, H.M, Hughes, M.S., and Hammond, J.M., "Test Facility Uncertainty Analyses for RBCC Systems Testing," AIAA-2002-3607, 38th AIAA/ASME/SAE/ASEE Joint Propulsion Conference, Indianapolis, IN, July 2002.

[3] Burner, A.W., Liu, T., and DeLoach, R., "Uncertainty of Videogrammetric Techniques used for Aerodynamic Testing, " AIAA 2002-2794, 22nd AIAA Advanced Measurement Technology and Ground Testing Conference, St. Louis, MO, June 24-26, 2002.

[4] DeLoach, R., "MDOE Perspectives on Wind Tunnel Testing Objectives, " AIAA 2002-2796, 22nd AIAA Advanced Measurement Technology and Ground Testing Conference, St. Louis, MO, June 24-26, 2002.

[5] Bartlett, E.K., "Evaluating the Design Process of a Four-bar-slider Mechanism using Uncertainty Techniques," M.S. Thesis, Dept. of Mechanical Engineering, Mississippi State University, Starkville, MS, May 2002.

[6] DeLoach, R., "Tactical Defenses Against Systematic Variation in Wind Tunnel Testing," AIAA 2002-0885, 40th AIAA Aerospace Sciences Meeting and Exhibit, Reno, NV, Jan. 14-17, 2002.

[7] Hudson, S.T. and Heng, B.L., "Evaluating Experimental Data Requirements for High Gradient Turbine Flowfields through Uncertainty Analysis," AIAA 2002-0883, 40th AIAA Aerospace Sciences Meeting and Exhibit, Reno, NV, Jan. 14-17, 2002.

[8] Heng, B.L., "Evaluating Data Averaging Techniques for High Gradient Flow Fields through Uncertainty Analysis," M.S. Thesis, Dept. of Mechanical Engineering, Mississippi State University, Starkville, MS, Aug. 2001.

[9] Liu, T., and Finley, T., "Estimate of Bias Error Distributions," AIAA-2001-0162, 39th AIAA Aerospace Sciences Meeting and Exhibit, Reno, NV, Jan. 8-11, 2001.

[10] Holman, J.P., *Experimental Methods for Engineers*, 7th Edition, McGraw Hill, 2001.

[11] Montgomery, D.C., Peck, E.A., and Vining, G.G., *Introduction to Linear Regression Analysis*, 3rd Edition, John Wiley and Sons, New York, 2001.

[12] Hudson, S.T. and Coleman, H.W., "Analytical and Experimental Assessment of Two Methods of Determining Turbine Efficiency," *Journal of Propulsion and Power*, Vol. 16, No. 5, Sept.-Oct. 2000, pp. 760-767.

[13] Belter, D., "Assessing Long-Term Instrumentation Performance and Uncertainty from Multiple Calibrations," 21st AIAA Aerodynamic Measurement Technology and Ground Testing Conference, Denver, CO, June 19-22, 2000.

[14] Hemsch, M. Grubb, J., Krieger, W., and Cler, D., "Langley Wind Tunnel Data Quality Assurance – Check Standard Results," AIAA-2000-2201, 21st AIAA Aerodynamic Measurement Technology and Ground Testing Conference, Denver, CO, June 19-22, 2000.

[15] Kammeyer, M. and Reuger, M., "On the Classification of Errors: Systematic, Random, and Replication Level," AIAA 2000-2203, 21st AIAA Aerodynamic Measurement Technology and Ground Testing Conference, Denver, CO, June 19-22, 2000.

[16] Possolo, A., Booker, A., and Witkowski, D., "Statistical Models and Error Analysis for Aerodynamic Variables," AIAA-2000-2696, 21st AIAA Aerodynamic Measurement Technology and Ground Testing Conference, Denver, CO, June 19-22, 2000.

[17] Wheeler, J., "Flight Test Applications of Uncertainty Analysis," AIAA-2000-2207, 21st AIAA Aerodynamic Measurement Technology and Ground Testing Conference, Denver, CO, June 19-22, 2000.

[18] DeLoach, R., "Improved Quality in Aerospace Testing through the Modern Design of Experiments (Invited)," AIAA-2000-0825, 38th AIAA Aerospace Sciences Meeting and Exhibit, Reno, NV, Jan. 10-13, 2000.

[19] Meyn, L., "An Uncertainty Propagation Methodology that Simplifies Uncertainty Analysis," AIAA-2000-0149, 38th AIAA Aerospace Sciences Meeting and Exhibit, Reno, NV, Jan. 10-13, 2000.

[20] Bendat and Piersol, *Random Data: Analysis and Measurement Procedures*, 3rd Edition, John Wiley & Sons, Inc., 2000.

[21] Mathcad 2001 Professional, Mathsoft Engineering and Education, Inc., 101 Main Street, Cambridge, MA, ©1986-2000.

[22] Springer, A.M., "Uncertainty Analysis of the NASA MSFC 14-inch Trisonic Wind Tunnel," AIAA-99-0684, 37th AIAA Aerospace Sciences Meeting and Exhibit, Reno, NV, Jan. 1999.

[23] "Assessment of Experimental Uncertainty with Application to Wind Tunnel Testing," AIAA Standard S-071A-1999.

[24] Coleman, H.W. and Steele, W.G., *Experimentation and Uncertainty Analysis for Engineers*, 2nd Edition, John Wiley & Sons, Inc., New York, NY, 1999.

[25] Brown, K.K., Coleman, H.W., and Steele, W.G., "A Methodology for Determining Experimental Uncertainties in Regressions," *Journal of Fluids Engineering*, Vol. 120, No. 3, Sept. 1998, pp. 445-456.

[26] Belter, D., "Comparison of Wind Tunnel Data Repeatability with Uncertainty Analysis Estimates," AIAA-98-2714, 20th AIAA Aerodynamic Measurement Technology and Ground Testing Conference, Albuquerque, NM, June 15-18, 1998.

[27] Hildebrandt, P. and Johann, E., "Reducing Static Pressure Measurement Errors to Increase Accuracy of Air Mass Flow Measurement," AIAA-98-2712, 20th AIAA Aerodynamic Measurement Technology and Ground Testing Conference, Albuquerque, NM, June 15-18, 1998.

[28] Hudson, S.T. and Coleman, H.W., "A Detailed Uncertainty Assessment of Methods used to Determine Turbine Efficiency," AIAA 98-2711, 20th AIAA Aerodynamic Measurement Technology and Ground Testing Conference, Albuquerque, NM, June 15-18, 1998.

[29] Kammeyer, M.E., "Wind Tunnel Facility Calibrations and Experimental Uncertainty," AIAA 98-2715, 20th AIAA Aerodynamic Measurement Technology and Ground Testing Conference, Albuquerque, NM, June 15-18, 1998.

[30] Mello, O.A.F., Uyeno, S., Sampaio, S., and Reis, M., "Uncertainty Methodology at the Brazilian TA-2 Subsonic Wind Tunnel," AIAA-98-2716, 20th AIAA Aerodynamic Measurement Technology and Ground Testing Conference, Albuquerque, NM, June 15-18, 1998.

[31] Meyn, L.A., "Software Tools for Measurement Uncertainty Analysis," AIAA-98-2713, 20th AIAA Aerodynamic Measurement Technology and Ground Testing Conference, Albuquerque, NM, June 15-18, 1998.

[32] Wilder, M. and Reda, D., "Uncertainty Analysis of the Liquid Crystal Coating Shear Vector Measurement Technique," AIAA-98-2717, 20th AIAA Aerodynamic Measurement Technology and Ground Testing Conference, Albuquerque, NM, June 15-18, 1998.

[33] Meyn, L.A., "A New Method for Integrating Uncertainty Analysis into Data Reduction Software." AIAA 98-0632, 36th AIAA Aerospace Sciences Meeting and Exhibit, Reno, NV, Jan. 12-15, 1998.

[34] Coleman, H.W. and Steele, W.G., "Uncertainty Analysis," Chapter 39, *CRC Handbook of Fluid Dynamics*, CRC Press, Inc., Boca Raton, FL, 1998.

[35] Draper, N.R. and Smith, H., *Applied Regression Analysis*, 3rd Edition, John Wiley and Sons, New York, 1998.

[36] Hudson, S.T., "Improved Turbine Efficiency Test Techniques Based on Uncertainty Analysis Application," Ph.D. Dissertation for the University of Alabama in Huntsville, Department of Mechanical and Aerospace Engineering, 1998.

[37] Steele, W.G. and Coleman, H.W., "Experimental Uncertainty Analysis," in Chapter 19, "Mathematics," *CRC Handbook of Mechanical Engineering*, CRC Press, Inc., Boca Raton, FL, 1998, pp. 19-118 – 19-124.

[38] Hosni, M.H., Coleman, H.W., and Steele, W.G., "Application of MathCAD Software in Performing Uncertainty Analysis Calculations to Facilitate Laboratory Instruction," *Computers in Education Journal*, Vol. 7, No. 4, Oct. - Dec. 1997, pp. 1-9.

[39] Hudson, S.T. and Coleman, H.W., "A Detailed Uncertainty Assessment of Measurements Used in Determining Turbine Efficiency," AIAA Paper 97-0776, 35th Aerospace Sciences Meeting & Exhibit, Reno, NV, Jan. 6-9, 1997.

[40] Martin, P. and Britcher, C.P., "Preliminary Flow Survey Measurements and Uncertainty Analysis from the Langley Full-Scale Tunnel," AIAA Student Paper, 1997 winner.

[41] Steele, W.G., Ferguson, R.A., Taylor, R.P., and Coleman, H.W., "Computer-Assisted Uncertainty Analysis," *Computer Applications in Engineering Education*, Vol. 5, No. 3, 1997, pp. 169-179.

[42] Markopolous, P., Coleman, H.W., and Hawk, C.W., "Uncertainty Assessment of Performance Evaluation Methods for Solar Thermal Absorber/Thruster Testing," *Journal of Propulsion and Power*, Vol. 13, No. 4, 1997, pp. 552-559.

[43] "American National Standard for Expressing Uncertainty – U.S. Guide to the Expression of Uncertainty in Measurement," ANSI/NCSL Z540-2-1997.

[44] Kammeyer, M., "Uncertainty Analysis for Force Testing in Production Wind Tunnels," NASA/CP-1999-209101/PT1, Proceedings of the First International Symposium on Strain-Gauge Balances, NASA Langley Research Center, Oct. 22-25, 1996, pp. 221-242.

[45] Hudson, S.T., Bordelon, W.J. Jr., and Coleman, H.W., "Effect of Correlated Precision Errors on Uncertainty of a Subsonic Venturi Calibration," *AIAA Journal*, Volume 34, Number 8, Sept. 1996.

[46] Steele, W.G., Maciejewski, P.K., James, C.A., Taylor, R.P., and Coleman, H.W., "Considering Asymmetric Systematic Uncertainties in the Determination of Experimental Uncertainty," *AIAA Journal*, Vol. 34, No. 7, July 1996, pp. 1458-1468.

[47] Belter, D.L., "Application of Uncertainty Methodology at the Boeing Aerodynamics Laboratory," AIAA Paper 96-2215, 19th AIAA Aerodynamic Measurement Technology and Ground Testing Conference, New Orleans, LA, June 17-20, 1996.

[48] Cahill, D.M., "Development of an Uncertainty Methodology for Multiple-Channel Instrumentation Systems," AIAA Paper 96-2216, 19th AIAA Aerodynamic Measurement Technology and Ground Testing Conference, New Orleans, LA, June 17-20, 1996.

[49] Everhart, J., "Calibration Improvements to Electrically-Scanned Pressure Systems and Preliminary Statistical Assessment," AIAA-96-2217, 19th AIAA Aerodynamic Measurement Technology and Ground Testing Conference, New Orleans, LA, June 17-20, 1996.

[50] Hedlund, E.R. and Kammeyer, M.E., "Aerodynamic and Aerothermal Instrumentation: Measurement Uncertainty in the NSWC Hypervelocity Wind Tunnel," 19th AIAA Aerodynamic Measurement Technology and Ground Testing Conference, New Orleans, LA, June 17-20, 1996.

[51] Hemsch, M., "Development and Status of Data Quality Assurance Program at NASA LARC—Towards National Standards," AIAA-96-2214, 19th AIAA Aerodynamic Measurement Technology and Ground Testing Conference, New Orleans, LA, June 17-20, 1996.

[52] Brown, K.K., Coleman, H.W., Steele, W.G., and Taylor, R.P., "Evaluation of Correlated Bias Approximations in Experimental Uncertainty Analysis," *AIAA Journal*, Volume 34, No. 5, May 1996, pp. 1013-1018.

[53] Hudson, S.T. and Coleman, H.W., "A Preliminary Assessment of Methods for Determining Turbine Efficiency," AIAA 96-0101, 34th Aerospace Sciences Meeting and Exhibit, Reno, NV, Jan. 15-18, 1996.

[54] Brown, K.K., "Assessment of the Experimental Uncertainty Associated with Regressions," Ph.D. Dissertation for the University of Alabama in Huntsville, Department of Mechanical and Aerospace Engineering, 1996.

[55] Coleman, H.W., Steele, W.G., and Taylor, R.P., "Implications of Correlated Bias Uncertainties in Single and Comparative Tests," *Journal of Fluids Engineering*, Vol. 117, Dec. 1995, pp. 552-556.

[56] Coleman, H.W. and Steele, W.G., "Engineering Application of Experimental Uncertainty Analysis," *AIAA Journal*, Vol. 33, No. 10, Oct. 1995, pp. 1888-1886.

[57] Balas, G.J., Lind, R., and Packard, A., "Robustness Analysis with Linear Time-Invariant and Time-Variant Real Uncertainty," AIAA 95-3188, Guidance, Navigation, and Control Conference, Baltimore, MD, Aug. 7-10, 1995.

[58] Wahls, R.A., Adcock, J.B., Witkowski, D.P., and Wright, F.L., "A Longitudinal Aerodynamic Data Repeatability Study for a Commercial Transport Model Test in the National Transonic Facility," NASA Technical Paper 3522, Aug. 1995.

[59] Blanton, J., "Uncertainty Estimates of Test Section Pressure and Velocity in the Large Cavitation Channel," AIAA-95-3079, 31st AIAA/ASME/SAE/ASEE Joint Propulsion Conference and Exhibit, San Diego, CA, July 10-12, 1995.

[60] Blumenthal, P.Z., "A PC Program for Estimating Measurement Uncertainty for Aeronautics Test Instrumentation," AIAA-95-3072, 31st AIAA/ASME/SAE/ASEE Joint Propulsion Conference and Exhibit, San Diego, CA, July 10-12, 1995.

[61] Greiner, B. and Frederick, R.A. Jr., "Labscale Hybrid Uncertainty Analysis," AIAA 95-3085, 31st AIAA/ASME/SAE/ASEE Joint Propulsion Conference, San Diego, CA, July 10-12, 1995.

[62] *Determining And Reporting Measurement Uncertainties*, NCSL Recommended Practice, RP-12, April 1995.

[63] Taylor, R.P., Luck, R., Hodge, B.K., and Steele, W.G., "Uncertainty Analysis of Diffuse-Gray Radiation Enclosure Problems," *Journal of Thermophysics and Heat Transfer*, Vol. 9, No. 1, Jan.-Mar. 1995, pp. 63-69.

[64] Brown, K.K. and Steele, W.G., "Estimating Uncertainty Intervals for Linear Regression," AIAA 95-0796, 33rd Aerospace Sciences Meeting, Reno, NV, Jan. 9-12, 1995.

[65] Taylor, R.P., Steele, W.G., and Douglas, F., "Uncertainty Analysis of Rocket Motor Thrust Measurements with Correlated Biases, " *ISA Transactions*, Vol. 34, 1995, pp.253-259.

[66] Taylor, B. and Kuyatt, C., "Guidelines for Evaluating and Expressing the Uncertainty of NIST Measurement Results," NIST Technical Note 1297, Sept. 1994.

[67] Cahill, D.M., "Experiences with Uncertainty Analysis Application in Wind Tunnel Testing," AIAA 94-2586, 18th AIAA Aerospace Ground Testing Conference, Colorado Springs, CO, June 20-23, 1994.

[68] Coleman, H.W., James, C.A., Maciejewski, P.K., Steele, W.G., and Taylor, R.P., "Considering Asymmetric Systematic Uncertainties in the Determination of Experimental Uncertainty," AIAA 94-2585 18th AIAA Aerospace Ground Testing Conference, Colorado Springs, CO, June 20-23, 1994.

[69] Tripp, J. and Tcheng, P., "Determination of Measurement Uncertainties of Multi-Component Wind Tunnel Balances," AIAA-94-2589, 18th AIAA Aerospace Ground Testing Conference, Colorado Springs, Colorado, June 20-23, 1994.

[70] Wright, F.L., "Comparison of Least Squares Curve Fit and Individual Sample Statistical Analysis Results of Calibration Data for the Velocity Coefficient of a Flow Nacelle," AIAA-94-2587, 18th AIAA Aerospace Ground Testing Conference, Colorado Springs, Colorado, June 20-23, 1994.

[71] Weiss, B. and Perry, L., "Uncertainty of Five-hole Probe Measurements in Water Tunnel Wake Surveys," ARL Technical Report No. IM 94-093, The Pennsylvania State University Applied Research Laboratory, June 7, 1994.

[72] Steele, W.G., Ferguson, R.A., Taylor, R.P., and Coleman, H.W., "Comparison of ANSI/ASME and ISO Models for Calculation of Uncertainty," Paper 94-1014, Proceedings of the 40th International Instrumentation Symposium, Sponsored by the Instrument Society of America, Baltimore, Maryland, May 1-5, 1994.

[73] Brown, K.K., Coleman, H.W., Steele, W.G., and Taylor, R.P., "Evaluation of Correlated Bias Approximations in Experimental Uncertainty Analysis," AIAA 94-0772, 32nd Aerospace Sciences Meeting & Exhibit, Reno, NV, Jan. 10-13, 1994.

[74] "Quality Assessment for Wind Tunnel Testing," AGARD-AR-304, 1994.

[75] Zierke, W., Straka, W., and Taylor, P., "The High Reynolds Number Flow Through An Axial-Flow Pump, (Appendix A: Five-Hole Probe Uncertainty Analysis)," ARL Technical Report No. TR 93-12, The Pennsylvania State University Applied Research Laboratory, Nov. 1993.

[76] Chakroun, W., Taylor, R.P., Steele, W.G., and Coleman, H.W., "Bias Error Using Ratios to Baseline Experiments – Heat Transfer Case Study," *Journal of Thermophysics and Heat Transfer*, Vol. 7, No. 4, Oct.-Dec. 1993, pp. 754-757.

[77] Steele, W.G., Taylor, R.P., Burrell, R.E., and Coleman, H.W., "Use of Previous Experience to Estimate Precision Uncertainty of Small Sample Experiments," *AIAA Journal*, Vol. 31, No. 10, Oct. 1993, pp. 1891 – 1896.

[78] Taylor, R.P., Hodge, B.K., and Steele, W.G., "Series Piping System Design Program with Uncertainty Analysis," *Heating/Piping/Air Conditioning*, Vol. 65, No. 5, May 1993, pp. 87-93.

[79] Steele, W.G., Taylor, R.P., Burrell, R.E., and Coleman, H.W. "Uncertainty of Small Sample Experiments," AIAA Paper 92-3954, 17th Aerospace Ground Testing Conference, Nashville, TN, July 6-8, 1992.

[80] Taylor, J., "Measurement Error – Its Nature and Origins," *Measurement and Controls*, June 1990.

[81] Moffat, R.J., "Describing the Uncertainties in Experimental Results," *Experimental Thermal and Fluid Science*, Vol. 1, Jan. 1988, pp. 3-17.

[82] Rood, E.P., Editor, "Experimental Uncertainty in Fluid Measurements," ASME Fluids Engineering Division Winter Meeting Session, FED-Vol. 58, Dec. 13-18, 1987.

[83] Patrick, W., "Flow Field Measurements and Reattached Flat Plate Turbulent Boundary Layer, (Appendix C: LV Error Analysis)," NASA Contract Report 4052, March 1987.

[84] Abernethy, R.B., Benedict, R.P., and Dowdell, R.B., "ASME Measurement Uncertainty," *J. Fluids Engineering*, Vol. 107, June 1985, pp. 161-164.

[85] Kline, S.J., "The Purposes of Uncertainty Analysis," *J. Fluids Engineering*, Vol. 107, June 1985, pp. 153-160.

[86] Lassahn, G.D., "Uncertainty Definition," *J. Fluids Engineering*, Vol. 107, June 1985, p. 179.

[87] Moffat, R.J., "Using Uncertainty Analysis in the Planning of an Experiment," *Journal of Fluids Engineering*, Vol. 107, June 1985, pp. 173-178.

[88] Taylor, J.R., *An Introduction To Error Analysis: The Study of Uncertainties in Physical Measurements*, University Science Books, Mill Valley, CA, 1982.

[89] Moffat, R.J., "Contributions to the Theory of Single-sample Uncertainty Analysis," *Journal of Fluids Engineering*, 104:250-260, June 1982.

[90] ICRPG Handbook for Estimating The Uncertainty in Measurements Made With Liquid Propellant Rocket Engine Systems, CPIA 180, April 1969.

[91] Grubbs, F.E., "Procedures for Detecting Outlying Observations in Samples," Technometrics, Vol. 11, No. 1, Feb. 1969.

[92] Ku, H.H., Editor, "Precision Measurement and Calibration: Selected NBS Papers on Statistical Concepts and Procedures," NBS Special Publication 300 – Vol. 1, Issued Feb. 1969.

[93] Natrella, M.G., "Experimental Statistics," National Bureau of Standards Handbook 91, Oct. 1966.

[94] Kline, S.J. and McClintock, F.A., "Describing Uncertainties in Single-Sample Experiments," *Mechanical Engineering*, Vol. 75, Jan. 1953, pp. 3-8.

[95] Berkson, J., "Are There Two Regressions," *Journal of the American Statistical Association*, June 1950, pp. 164-180.

[96] Thompson, W.R., "On a Criterion for the Rejection of Observations and the Distribution of the Ratio of the Deviations to Sample Standard Deviation," *Annals of Mathematical Statistics*, Vol. 6, 1935, pp. 214-219.

[97] Dieck, R.H., "Measurement Uncertainty: Methods and Applications," 3rd Edition, The Instrumentation, Systems, and Automation Society, Product ID/ISBN 1-55617-795X.

[98] Eisenhart, C., "Expression of the Uncertainties of Final Results," NBS Special Publication 300, Vol. 1, pp. 69-72.

[99] Ku, H.H., "Expressions of Imprecision, Systematic Error, and Uncertainty Associated with a Reported Value," NBS Special Publication 300, Vol. 1, pp. 73-78.

[100] Youden, W.J., "Uncertainties in Calibration," NBS Special Publication 300, Vol. 1, pp. 63-68.

[101] Youden, W.J., "Systematic Errors in Physical Constants," NBS Special Publication 300, Vol. 1, pp. 56-62.